Astronomers' Observing Guides

Series Editor
Dr. Michael D. Inglis, BSc, MSc, Ph.D.
Fellow of the Royal Astronomical Society
Suffolk County Community College
New York, USA
inglism@sunysuffolk.edu

For further volumes:
http://www.springer.com/series/5338

Brian Cudnik

Faint Objects
and How to
Observe Them

with 69 Illustrations

 Springer

Brian Cudnik
Leaf Oak Drive 11851
Houston, Texas, USA

ISSN 1611-7360
ISBN 978-1-4419-6756-5   ISBN 978-1-4419-6757-2 (eBook)
DOI 10.1007/978-1-4419-6757-2
Springer New York Heidelberg Dordrecht London

Library of Congress Control Number: 2012942846

Printed on acid-free paper

Springer is part of Springer Science+Business Media (www.springer.com)

I would like to dedicate this to my astronomy friends who have passed on from this life: Don "Captain Comet" Pearce, Rick Hillier, George Stradley, Richard Bunkley and others. May your legacies shine on like stars in the universe.

About the Author

Brian Cudnik has been an amateur astronomer for over 30 years and manages the Physics laboratories at Prairie View A & M University (a part of the A & M University of Texas). He has been the coordinator of the Lunar Meteoritic Impact Search section of the Association of Lunar and Planetary Observers (ALPO) since January 2000. Cudnik began at ALPO 2 months after it made the first confirmed visual observation of a meteoroid impact on the Moon during the Leonid storm of November 1999. Cudnik has an M.Sc. and has published papers and posters on various astronomical subjects, both peer-reviewed and amateur. He has served as a board member of the Houston Astronomical Society, is presently an Associate member of the American Astronomical Society, a member of the American Meteorological Society, a member of the American Association of Physics Teachers, and a regular contributor of observations to the American Association of Variable Star Observers and the International Occultation Timing Association. He teaches astronomy at the University of St. Thomas two evenings per week each semester.

Preface

Pushing the Envelope in Visual Astronomical Observations

Astronomy encompasses an unimaginably vast and complex universe of objects, from the planets, moons, comets, and asteroids of the local Solar System to the most distant galaxies— and everything in between. Most of what we can see easily or are prone to look at through the eyepiece in the nighttime sky are the nearest and/or the brightest representations of the various astronomical objects. Most amateur astronomers who actively observe have seen the likes of M42, the Great Orion Nebula, or M57, and the Ring Nebula. But how many people have seen PK 013.3 + 32.7? PK 013.3 + 32.7 is also known as Shane 1 and I had not even heard of this object until the summer of 2010, when it was mentioned in the June 2010 issue of *Sky & Telescope* (p. 65 of that issue). Bright, nearby objects are always fun or rewarding to look at, but one can truly expand one's cosmic horizons by hunting down obscure deep sky objects and seeing things that few others have seen.

I have included my own drawings of the Great Orion Nebula, along with the planetary nebulae NGC 6772 and NGC 6872 (which people may or may not have seen, but both appear as very different shapes through the eyepiece) below in Fig. i.1

This book, *Faint Objects and How to Observe Them*, primarily deals with techniques in visual astronomy to enable one to observe these "elusive faint fuzzies." The visual side of astronomy is an art form as well as a science in that a full appreciation of the subtle beauty of astronomical objects takes time to develop. For amateur astronomers who have not done deep astronomy very long, I recommend starting with the brightest, most spectacular objects (and revisit them regularly) like the Moon, Saturn, Jupiter, the Orion Nebula, the Andromeda Galaxy and many others, and continue your pursuits with the brighter deep sky objects. The vast majority of deep sky objects (at least the ones that we can see from Earth) are quite faint, which is the primary focus of this book.

The target audience is the intermediate to advanced visual observer. If you are a beginner who wants to "go deep," you can also benefit from this book; however, it is recommended that you master the ability to locate the brighter Messier and NGC objects before attempting the fainter objects in this book. We will skip over the most basic elements of observing techniques (although there are some refresher tidbits on the basics from time to time) and focus mainly on the skills needed to locate, observe, and appreciate faint objects. Figure i.2 show just how deep the

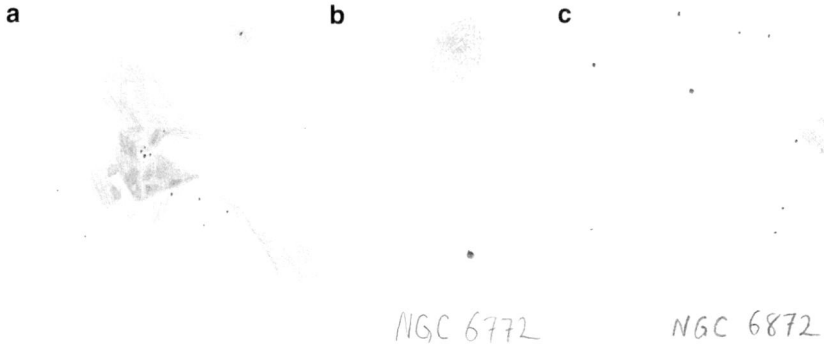

Fig. i.1 Author's drawings of (**a**) M42, (**b**) NGC 6772, and (**c**) NGC 6872. M42 was observed with an 8-in. f/10 Cassegrain at 83× and 226× and under light pollution; NGC 6872 was observed under dark skies, with an OIII filter and a 14-in. f/11 Cassegrain (referred to as the "C14" here and elsewhere in this book) at 244.5×; and NGC 6772 was sighted with the C14, no filter, at 98×

Fig. i.2 Author's drawing of the Ursa Major double quasar, visual observation from the 2003 Texas Star Party, with Larry Mitchell's 36-in. telescope at 300×

author has observed; it is a drawing of the Ursa Major Double Quasar observed during the 2003 Texas Star Party through the eyepiece of a 36-in. (91-cm) reflecting telescope. This object lies some 7 billion light years from Earth.

The term "faint objects" can include a wide variety of objects, from atmospheric phenomena such as sprites (bursts of light above thunderstorms) and other atmospheric transients to Solar. System objects such as faint asteroids, comets and distant dwarf planets, and to objects outside the Solar System such as nebulae, clusters, individual galaxies and galaxy clusters. This book will concentrate on deep sky objects and leave the Solar System objects, however faint, to other books that are dedicated to a particular type of object (for example, faint comets described in the book *Comets and How to Observe Them*, another book in the Springer "Astronomers' Observing Guides" series).

The following historical passage sheds a little light on the discovery of the nature of faint objects as well as the uncovering of even more faint objects. In fact, it was not until the third decade of the twentieth century that we truly became aware of the fact that the Milky Way Galaxy is not the entire universe but rather one of many galaxies in a universe far larger and more complex than we ever imagined up to that point. With larger and larger telescopes being built and the quality of the optics improving in the first decades of the twentieth century, our view of the cosmos became more and more clear. Examples of these large telescopes include the 100-in. (2.54-m) telescope at Mt. Wilson, which became operational in 1917 (Fig. i.3a), and the 200-in. (5.08-m) Palomar Hale 'scope that received its first light in 1949 (Fig. i.3b) and remained the largest telescope in the world until 1976. At this time the BTA-6 telescope in Russia became operational as the world's largest, and would remain so until the Keck I telescope (Fig. i.3c), at 387 in. (10 m) became the world's largest telescope in operation in 1993. With these larger telescopes and the wider and more effective use of photography, nearby galaxies, for the first time, became resolved into individual stars.

Galaxies, the so-called "spiral nebulae" of the eighteenth and nineteenth centuries, were noted during the surveys that will be described in more detail in Chap. 1. They were thought to be planetary systems in the making, all within our own galaxy. In 1920, two astronomers, Harlow Shapley and Heber D. Curtis, began what would later be called the Shapley-Curtis debate. Curtis argued that the universe was much larger than had been accepted up to this point, being composed of many galaxies (the "spiral nebulae") like the Milky Way. But Shapley argued that these objects were actually nearby gas clouds, and that the universe was made of only one great big galaxy. The Sun, in Curtis's cosmogony, was near the center of our relatively small galaxy; Shapley contended that the Sun was far from the center of the galaxy.

The realization by the mid 1920s that the Andromeda Galaxy is actually made up of numerous stars, along with other discoveries in the 1930s, partially resolved this debate. Edwin Hubble was instrumental in the first discovery, as special types of variable stars called Cepheids were discovered in Andromeda. This, combined with the work of Henrietta Leavitt (an American astronomer who discovered the period-luminosity relationship of these objects), made it possible to find the distance to Cepheid variables in Andromeda, and by extension, Andromeda itself. Hubble's work also revealed that galaxies were receding from us at speeds proportional to their distances, and the expansion of the universe was discovered (this discovery led to the Big Bang theory of the origin of the universe which, as of this writing, remains the leading explanation of the origins of the Cosmos; our changing view of the universe over the last millennium is illustrated in Fig. i.4).

More on the Shapley-Curtis debate: it seems that both astronomers were correct to an extent. Shapley was more correct about the size of the Milky Way Galaxy and the location of the Sun within it, and Curtis was correct that the universe contains

Fig. i.3 (**a**) Images of the Mt. Wilson Observatory, looking toward the 100-in. telescope (aerial view, the larger dome one), (**b**) Ground image of the dome housing the 200-in. Hale telescope at Palomar observatory (Image courtesy of Mr. Scottthezombie of Wikimedia), and (**c**) the twin domes of the 8-m Keck telescopes (Image courtesy of NASA)

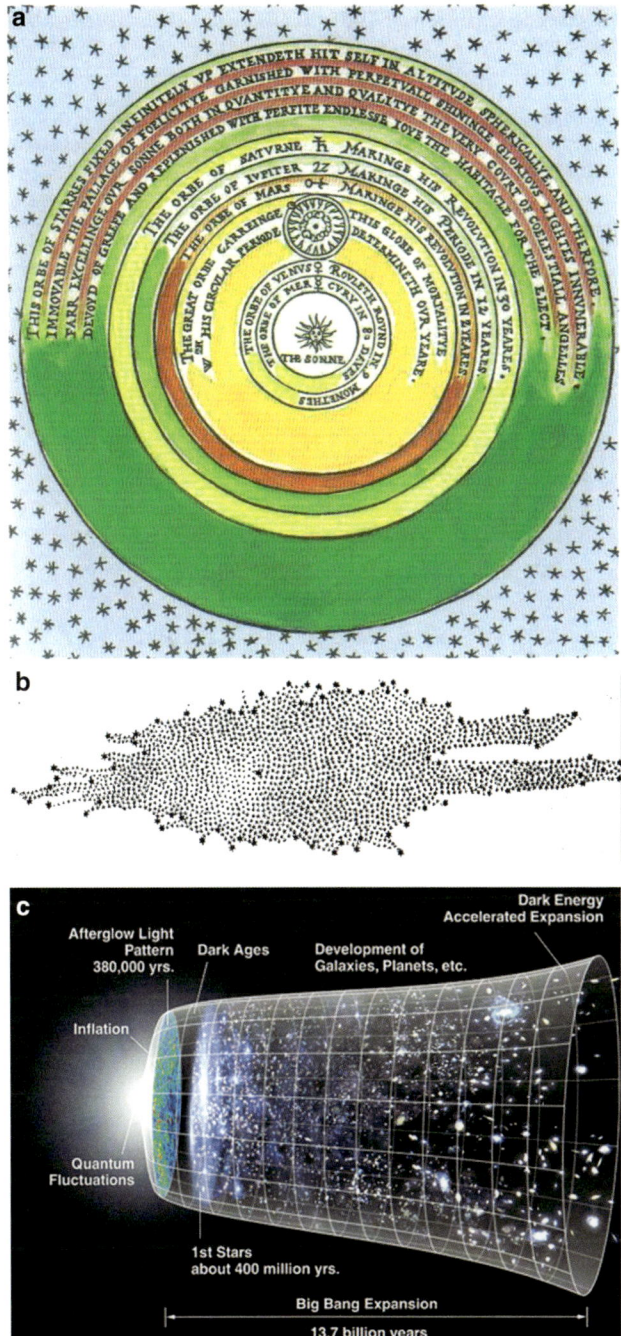

Fig. i.4 Representative images of our view of the universe over the years: (**a**) a model of the Copernican universe by Thomas Digges in 1576, who states that the stars are spread uniformly throughout space outside the realm of the planets (they are not confined to a sphere), (**b**) William Herschel's model of the Milky Way Galaxy based on his observations and research, and (**c**) Early twenty-first-century view of the universe in space and time (All images courtesy of Wikimedia)

many more galaxies than our own and that "spiral nebulae" are actually galaxies just like our own.

Back to the telescopes: as 'scopes became larger and larger and techniques and technologies vastly improved (some examples of these are the advent of electronic detectors, the deployment of the Hubble Space Telescope in orbit around Earth, and the use of adaptive optics and interferometry in optical astronomy), the number of known objects, the variety of objects, the dynamics of the universe, and astronomical knowledge in general grew exponentially. With all this development came wider availability of larger telescopes (in the 14-in./35-cm and larger class) to the general public, with monsters 50 in. (1.28 ms) for those with lots of money (It costs the same as a small house in the United States!).

With the developments described above, and the availability of affordable electronic equipment and excellent techniques to the general public, amateur astronomers are producing results comparable to professionals just a few "short" decades ago. In effect, equipped with larger telescopes, better optics, and enhanced observing techniques, amateurs are able to "see farther" than ever before. With these advances, more and more of the universe has become available to the general public. Unfortunately more and more of the universe is becoming hidden by expanding domes of light pollution blanketing the world's cities, rendering faint objects invisible or difficult to see at best. This increase in light pollution that comes with development threatens to make observing faint objects visually more and more difficult as we move deeper into the twenty-first century.

Having shared all this, we ask the question: "Why observe faint objects at all?"

Finding and observing faint objects bring a certain level of satisfaction to those who pursue them for these reasons, most of which will be explored in more detail throughout the book:

- Faint objects gives the observer an opportunity to see what few amateurs get to see (either through their own effort or through other people's 'scopes).
- Faint objects give the observer the opportunity to see deeper into the universe than their colleagues.
- Faint objects greatly increases the inventory of objects that one has seen, to include more unique objects (such as quasars and BL Lacertae objects) than many amateur astronomers have seen in their entire observing "careers".
- Faint objects provide a measure of satisfaction of hunting and viewing challenging objects (the thrill of the hunt).
- Faint objects enable the amateur astronomer to hone and master fine observing skills. These fine skills will help the same individual better see fine details in brighter objects as well as the improved ability to pick out other types of faint objects such as comets.

There are other reasons that astronomers (including you, the reader) who like to go deep would probably include here in addition to these listed.

The bottom line is that, in many cases, observing faint objects connects the astronomer with light that has traveled through interstellar or intergalactic space for up to hundreds of millions or even billions of years. As a result, we are witnessing directly the history of the universe: we see objects as they appear thousands, millions, hundreds of millions or even billions of years ago. So astronomy not only involves a look at distant objects in the universe but also a look back in time, at the history of this same universe.

In the pages that follow, we will sample some challenging open and globular star clusters, planetary nebulae, emission and reflection nebulae, and galaxies. A few quasars will be thrown in for good measure as well. We will cover distant parts of our galaxy as well as distant parts of the universe. A few nearby but small objects will be included also, such as dwarf galaxies and low surface brightness planetary nebulae.

Part I of the book covers the physical nature of these selected objects. What are star clusters? What are planetary nebulae? What is the difference between an emission and a planetary nebula? How are surveys conducted, and how have these objects been discovered? Who has been instrumental in bringing these objects into the realm of the known universe? These will be investigated in the first half of the book. Toward the end of the first half, we will consider why the objects are so faint to begin with and what the significance of their faintness is. We also consider how all these objects may be related to one another. The first half of the book ends with a brief overview of some of the cutting-edge observatories and surveys that continue to hunt for ever fainter and farther objects.

Part II goes into detail about the practical side, the "how to" of observing faint objects. The requirements or prerequisites for viewing such objects are laid out, to include a telescope with at least a 10-in. aperture; access to very dark skies; choosing the best, most transparent and moonless nights to go observing; the proper use of filters; and the maintenance of the equipment. These are described in more detail in Chap. 7, which includes in-depth discussions of the above ideas, plus observation planning that maximizes what often is precious little dark sky time. A balance between quantity and quality is recommended; that is, viewing significant numbers of objects but taking the time to enjoy the uniqueness of each object. Near the end, in Appendix A, we will also feature some software programs such as *Sky Tools* that assist in planning, executing, and documenting an observing session.

After having selected the site, prepped the 'scope, made the observing list (whether with pencil and paper or with the aid of your favorite software), and secured favorable weather with the Moon out of the way, now is the time to get down to some serious observing business. A selection of projects to get you started will be listed, beginning with the projects that the Astronomical League provides for intermediate observers. The "Herschel 400" list is a great place to start and provides a warm up for even fainter objects. Then object types will be broken down by discrete groups (nebulae, clusters, galaxies) and groups of objects themselves, such as Hickson and Arp, and links to information about these and other surveys will be provided. Some suggestions on how to best record what one sees will be given. We do not stay in the Milky Way but include nebulae and star clusters that can be seen in other galaxies.

There are literally hundreds of thousands of objects (the vast majority of which are galaxies) available for observation to one equipped with, say, a 17.5-in. (44.5-cm) telescope; that number swells to the millions (again, almost all galaxies) for people with 36-in. (91-cm, and larger) 'scopes. Although it is nearly physically impossible for an individual astronomer, by himself or herself, to observe all of these objects visually, one can try for as many as one can get and appreciate the ones that he/she does see. In Chap. 10 is a catalog of over 700 representative objects as well as information on where to find additional lists (this in Appendix C), and share the author's personal observations of a sample of these objects.

Finally we explore some ways that, once having started your quest for faint objects, you can stick with this and help you to keep going for the long haul. This is covered in Chaps. 11 and 12, along with some ways that you can make your observations scientifically useful. Resources available to help find variable stars and supernovae are provided and how to estimate their magnitudes. Also described is "Citizen Science" and many ways that the reader can help with actual scientific research, from making careful visual observations to using your PC's idle time to help analyze lots and lots of data. Things wrap up with a conclusion that puts all of this together and outlines a typical observing run (at least what is typically encountered, which encompasses several areas of astronomy, including faint objects).

Acknowledgments

I would like to thank the following people, without whom this project would not be possible. It is their combined effort and my assembling their input that has made this observer's guide the best that it could be.

My wife Susan for her companionship and encouragement throughout the completion of this book.

Steve Gottlieb for providing most of the information for the faint objects observing lists in Chap. 9.

Don Taylor of the Houston Astronomical Society for allowing me to use his spectacular images.

Paul and Liz Downing for allowing me to use several of their beautiful images.

Loyd Overcash for allowing me to use many of his wonderful images.

Clayton Jeter, Ed Fraini, Ken Drake, Greg Barolak, Terrence Redding, Rodger Jones, Chris Ober, Larry Wadle, Tony George, and Dick Lock for their useful observing software feedback.

The many people of the "amastro" Amateur Astronomy Mailing List, who provided many useful ideas to get me started on the manuscript.

The many people, such as Mike Oates, Edmund Robertson (University of St. Andrews in Scotland), Dan Lewis (Huntington Library), Hartmut Frommert (SEDS), Michael Saladyga (AAVSO), Gary Kronk, Bob Argyle and John Isles (the Webb Deep Sky Society), Robert Naeye (editor), *Sky & Telescope*, Camille Carlisle (*Sky & Telescope*), Barbara Fraps (National Optical Astronomy Observatory), and Caroll Iorg (president of the Astronomical League) who granted me permission to use their stuff.

Maury Solomon, editor, Springer (U.S.) for her guidance and assistance with this project.

John Watson for providing an up-to-date author's guide and his guidance.

Greg Crinklaw for allowing me to use information about Sky Tools to enable this to be featured as a recommended product for the observations of faint objects.

The many members of the Houston Astronomical Society, who offered their encouragement and suggestions, and the club for ongoing access to their Dark Site Observing Site and resident telescopes.

Anyone else who I may have failed to mention, including those who gave permission to use their material in this book, those who gave advice on what to include, and those who gave encouragement and motivation to complete this work over the lifetime of the project.

Contents

Part II How to Observe Faint Objects

Contents

Part I

The Physical Nature of Faint Objects

The Astronomical Surveys

Historical Perspective

Astronomy is the oldest of the physical sciences and can be traced back to antiquity or at least 5,000 years ago [1]. Looking that far back in time, we see the origins of astronomy mixed in with the religious and astrological practices of pre-history. Ancient societies were already able to tell the wandering planets from the "fixed stars" and associated these moving objects with gods and spirits. Various cultures have practiced various religions related to astronomy, with the priests playing the role of the "professional astronomer" of their time, and demonstrating a "divine" understanding of the movements of the heavens. So it seems that the first few millennia of astronomical involvement of human societies were religious in nature.

Astronomy has also served a time keeping purpose for the majority of history. The basic motions of Earth and the Moon have, for most (if not all) of recorded history, defined our days, months, and years. Agricultural societies made use of the stars and star patterns as a sort of calendar to know when it was time to plant and when it was time to harvest. One example that comes to mind is the first appearance of the bright star Sirius in the dawn (in August for the mid-northern latitudes nowadays) as noted by the ancient Egyptians. They used this to know when the Nile River would flood and to serve as the start of their calendar year. Due to precession, this occurred near the summer solstice, the longest day of the year for the ancient Egyptians. Western astronomy has its origins in Mesopotamia with the ancient kingdoms of Assyria, Babylon, and Sumer. The earliest known star catalogs originated in Babylonia around 1200 B.C. The Babylonians were among the first (as far as we know) to recognize the periodic nature of astronomical phenomena.

Since the time of the Babylonians, and leading up to the Greek astronomer Hipparchus, people have been interested in surveying the natural world and cataloging what they find. In fact, since the invention of the telescope, and even before, astronomers have been surveying the skies, from Hipparchus cataloging stars in ancient Greece to Messier observing telescopic fuzzy objects from Renaissance France. Astronomers investigated the heavens just as explorers surveyed the surface of Earth. In both cases, new territory was to be found, cataloged, classified, and further studied. Astronomers were limited to what the naked eye could see prior to 1609, when the telescope was first used to look at the skies. Approximately 5,000

B. Cudnik, *Faint Objects and How to Observe Them*, Astronomers' Observing Guides,
DOI 10.1007/978-1-4419-6757-2_1, © Springer Science+Business Media New York 2013

stars, the Sun, the Moon, and five naked-eye planets rounded out the pre-telescopic inventory of the cosmos. A handful of deep-sky objects, as well as the occasional comet and supernova, also added to the listing of objects in the known universe prior to the astronomical use of the telescope.

Pre-Telescopic Discoveries of Deep Sky Objects

Deep sky objects known since well before the use of optical aid are listed as follows. The Pleiades and the Hyades were both known to the ancient Greeks and were included in their mythologies; the Pleiades were known pre-historically and were mentioned by Homer about 750 B.C. and by Hesiod about 700 B.C.[2]. The Beehive star cluster, also known as Praesepe (and also M44), was first cataloged by Ptolemy, making it the second earliest cataloged deep sky object and was also observed as early as 260 B.C. by Aratos [3]. Another bright and familiar object, M31 was known to Persian astronomer Abd-Al-Rhaman Al-Sufi (who referred to the object as the "little cloud") around A.D. 905 or 954 [4]. Its true nature would not be revealed for another thousand years. The Great Orion Nebula, M42, was probably discovered in 1610 by Nicholas-Claude Fabri de Peiresc. It was also independently discovered by Cysatus in 1611 [5].

More examples of pre-telescopic discoveries of deep sky objects include M7, Ptolemy's cluster known as early as A.D. 130, and Ptolemy himself has described this as "the nebula following the sting of Scorpius." [6] It is possible that M39, the open cluster in Cygnus, was observed by Aristotle in 325 B.C., who noted it as a "cometary appearing object" [7]; he possibly saw M41 as well in 325 B.C. The double cluster, not cataloged by Messier since he was mainly interested in comet-like objects, was likely cataloged 130 B.C. by Hipparchus [8]. Finally, the Coma Star cluster, Melotte 111, used to be Leo's tail, but Ptolemy III renamed it in 240 B.C. for the Egyptian Queen Bernice's sacrifice of her hair, as described by legend (Fig. 1.1 includes modern-day images of three of these objects: M31 (a), M44 (b), and the Double Cluster (c))) [9].

Two Short Observing Projects Related to These Objects

Although observing projects as a whole aren't introduced until Chap. 9 of this book, one project recommended to interested readers is to find each of the objects in Table 1.1 with the naked eye. These may not necessarily be "faint objects" per se, but they give the opportunity to retrace the footsteps (eye-steps?) of those who paved the way in our understanding of what is out there in general over the course of several millennia. To get the best results (that is, views that most closely approximate what these pioneering astronomers must have seen), use an instrument that closely matches the aperture (of the mirror or lens) of that used by each astronomer, or use your own if you do not have access to such telescopes.

Another interesting project for modern amateur astronomers, especially if they are just beginning deep sky observations, is to give them a list of object names and celestial coordinates and ask them to come up with a classification scheme with which to categorize each object on the list. In so doing, he or she is recreating the

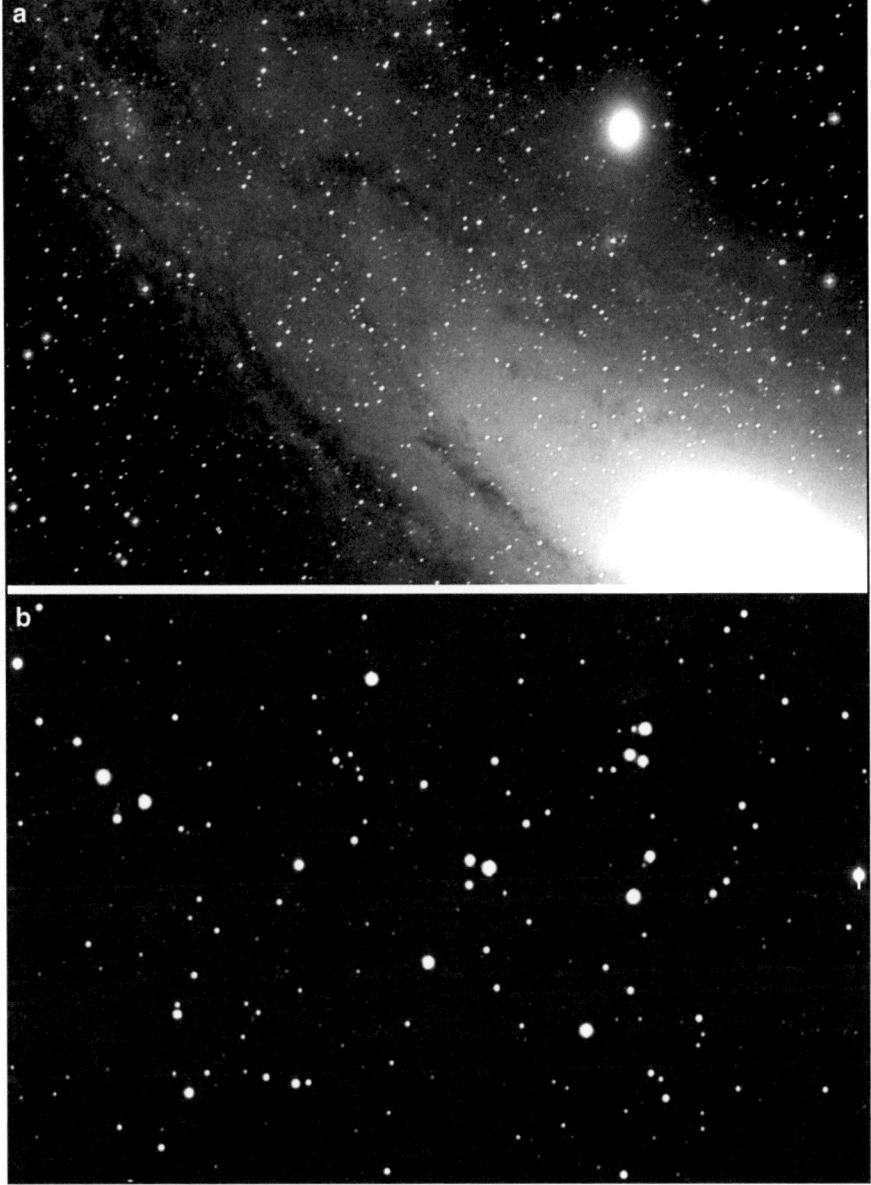

Fig. 1.1. Examples of objects known before the invention of the telescope: (**a**) M31, the Andromeda Galaxy, and (**b**) M44, the Beehive Cluster; and (**c**) the Double Star Cluster (All three images are courtesy of Paul and Liz Downing).

Fig. 1.1. (continued)

Table 1.1. Years of discovery (or at least when they were first noted and recorded and the record survives) of some of the brighter deep sky objects

Object	Discoverer/Person who first noted object(s)	When discovered or first noted
Pleiades (M45)	Homer/Hesiod	750 B.C./700 B.C.
M39 open cluster	Aristotle	325 B.C.
M41 open cluster	Aristotle	325 B.C.
Beehive/Praesepe (M44)	Aratos	260 B.C.
Melotte 111 (Coma star cluster)	Ptolemy	138 B.C.
M7 open cluster	Ptolemy	130 B.C.
Double Cluster	Hipparchus	130 B.C.
Andromeda Galaxy (M31)	Abd-Al-Rhaman Al Sufi	905 A.D.
Orion Nebula (M42)	Nicholas-Claude Fabri de Peiresc/Cysatus	1610, 1611
Clusters M6, IC 4665, NGC 6633, M26, M25, and M35	Philippe Loys de Chéseaux	1745
Crab Nebula (M1)	John Bevis	1731
M2 (Globular Cluster)	Jean-Dominique Maraldi	Sep. 1746
NGC 2360	Caroline Herschel	Feb. 1783
NGC 205	Caroline Herschel	Aug. 1783
NGC 253, NGC 381, NGC 659, NGC 2548 (M48), NGC 6633, NGC 7789	Caroline Herschel	Fall 1783

first century or two of telescope-aided astronomy as various astronomers of the day surveyed the skies and attempted to classify the many new non-stellar objects that were being discovered at the time. Some of the observing clubs of the Astronomical League ask participants to do just that, but within specific object types rather than across the board (more on these projects in Chap. 9).

Categorizing and Cataloging Deep Sky Objects

A Brief History

By the end of the first millennium A.D. a handful of non-stellar objects was known, most of which were true star clusters and nebulae, but many were asterisms or "false" clusters [10, 11]. These and other "false" clusters were included in the star catalogs of Tycho Brahe and Ulugh Begh. With the use of the telescope by Galileo Galilei, many of the unresolved nebulae were broken down into individual "tiny" stars. It was thought by Galileo and many others over the seventeenth and eighteenth centuries that all nebulae could be resolved into clusters of stars. But as telescopes and techniques improved over this time, it became clear that some of the objects were star clusters and some were true nebulae. The French astronomer de Chésaux discovered the following objects: M6, IC 4665, NGC 6633, M16, M25, and M35. He also more or less successfully distinguished between nebulae that were actually clusters and true nebula, which appeared as gray clouds.

The Orion Nebula, though visible with the naked eye and at least as prominent as the Andromeda Galaxy in the night sky, was not discovered until November 26, 1610 [12]. This was when Nicolas-Claude Fabri de Peiresc found it using a telescope. The nebula was also seen by Johann Baptist Cysat in 1618 but would not be studied in detail until 1659, when Christian Huygens observed it (he thought he was the first one to discover the nebula). Over the years, the number of nebulae that were known grew. Edmund Halley published a list of 6 nebulae in 1715, and Jean-Philippe de Cheseaux assembled a list of 20 nebulae (including 8 not previously known) in 1746.

As telescopic observations became more and more commonplace, and the numbers of new and diverse nebulous objects known to astronomers grew dramatically, one of the challenges was to successfully classify the myriad of objects that were discovered. In most cases, it was easy to tell between an open cluster, a globular cluster, and a nebula, but the lines of distinction were not always clear between "adjacent" kinds of objects. The astronomer Abbe Lacaille made an expedition to the Cape of Good Hope in 1751–1753 and made a list of 42 nebulous objects observed in the southern skies (most previously unknown). He also attempted a more rigorous classification scheme of the "nebulae" that he found, which went like this:

- Class I—nebulae without stars
- Class II—nebulous stars in clusters
- Class III—stars accompanied by nebulosity

This scheme did make a rough distinction between open star clusters, globular star clusters, and diffuse nebulae. However, the distinction was not sharp enough or consistent enough to be meaningful at the time.

Globular clusters were not known until a German amateur astronomer by the name of Abraham Ihle discovered M22 in 1665 [13]. Because of the small apertures and less-than-ideal optics of the early telescopes, globular clusters were not resolved into individual stars and were even considered a different type of nebula until Charles Messier's observation of M4. The next globular to be discovered was omega (ω) Centauri by Edmond Halley in 1677. Halley also discovered M13, the fourth such object to be found, in 1714. Gottfried Kirch found the third known globular, M5, in 1702. Philippe Loys de Chéseaux (of de Chéseaux comet fame) found M71 and M4 in 1745 and 1746, respectively. Finally, in 1746, Jean-Dominique Maraldi discovered the seventh and eighth known globulars M15 and M2.

From there the list grew more rapidly, with Abbé Lacaille listing NGC 104, NGC 4833, M55, M69, and NGC 6397 in his catalog generated in 1751–1752. William Herschel began his survey in 1782, and with larger telescopes, he was able to resolve stars in all of the 33 known globulars of the time. He added to that number, finding 37 additional globulars; he was the first to use the term "globular cluster" in his 1789 catalog of deep sky objects. Astronomers continued to add to the number of globular clusters over the next 125 years, reaching 83 objects by 1915. That number grew to 93 by 1930, and 97 by 1947. Today there are between 152 and 160 (depending on the source of the census) globular clusters in the Milky Way, out of an estimated grand total of 180 ± 20 globulars. Most of these undiscovered globulars are likely hidden behind clouds of gas, dust, and stars of the Milky Way.

Charles Messier, one of the first to do a systematic survey and catalog of the deep sky objects (as they are called today), did not attempt to classify the nebulae, but he did discover a number of new objects (more information on his life and work appear in the next chapter). He compiled a catalog that included a total of 103 objects, many of which are what we now know as galaxies. It was William Herschel who classified his discoveries into eight categories:

- Bright nebulae
- Faint nebulae
- Very faint nebulae
- Planetary nebulae
- Very large nebulae
- Very compressed and rich clusters of stars
- Compressed clusters of small and large (faint and bright) stars
- Coarsely scattered clusters of stars

Herschel was thought to have been the first person who coined the term "planetary nebulae" (object type IV) to describe these objects, back in 1785, but there is a reference to Darquier in 1781 who described the Ring Nebulae (M57, which he discovered) as "looking like a fading planet" [11]. These objects continued to be discovered and observed until the rise of spectroscopy in the 1860s revolutionized our knowledge of these objects. Huggins studied the planetary nebula NGC 6543 spectroscopically and discovered emission spectral lines, which set this object (and others like it) apart from clusters of stars that appeared nebulous. This led to the realization that not all nebulae are unresolved clusters of stars and dispel the belief that had been held since the discovery of the telescope.

An emission line spectrum consists of bright lines against a faint continuum or dark background; an absorption line spectrum (characteristic of most stars and star clusters) is made up of a rainbow of colors (a bright continuum) with thin dark lines superimposed. This was first discovered with the Sun and its famous Fraunhofer spectrum. It was believed that the element that made up nebulae was a new element that was called nebulium and was thought only to be found in gaseous nebulae. It was later realized, in the 1920s, that nebulium was actually twice-ionized oxygen (oxygen that had lost two of its electrons, written as [OIII], the brackets signifying that the spectral line from this ion is a "forbidden" line—it cannot be readily reproduced in an Earth-based laboratory).

With this new tool, astronomers were able to begin classifying objects by more than their visual appearance. In fact, this tool enabled nebulous objects that would otherwise appear stellar to be discovered and identified as such. Since the spectral nature of planetary nebulae was discovered just before Dreyer compiled the New General Catalogue, almost all of the NGC objects are extended objects, but many of the Index Catalogue nebulae (as well as nebulae discovered during subsequent surveys) appear stellar or almost stellar.

As a result of these surveys, many catalogs listing deep sky objects of all types have been assembled over the years, some of which are listed immediately below. Some catalogs focus on specific types of object, while others are more general in their coverage, and there is a lot of overlap between catalogs. A sample of catalogs currently available includes the following [14] (those described in this book in at least some detail are listed in bold font):

Abell—Abell Catalog
C—Caldwell Catalog
Col—Collinder Catalog
FCC—Fornax Cluster Catalog
FSC—Faint Source Catalog
GC—General Catalog of Nebulae and Clusters
IC—Index Catalog (IC I—Index Catalog I; IC II—Index Catalog II)
LBN—Lynds' Catalog of Bright Nebulae
LEDA—Lyon-Meudon Extragalactic Database
NGC—New General Catalogue
PGC—Principal Galaxies Catalogue
PK—Catalogue of Galactic Planetary Nebulae (Perek-Kohoutek)
PNG—Strasbourg-ESO Catalog of Galactic Planetary Nebulae
QSO—Revised and Updated Catalog of Quasi-stellar Objects
RNGC—Revised New General Catalogue
RSA—Revised Shapley-Ames Catalog
Sh—Sharpless Catalog (Sh 1 [1953] & Sh 2 [1959])
Stock—Stock open clusters
UGC—Uppsala General Catalogue
VCC—Virgo Cluster Catalog
Z—Fritz Zwicky, Catalog of galaxies and of clusters of galaxies

Early Catalogs

The first deep sky catalog with a significant number of galaxies, conducted with telescopes, was Charles Messier's, the observations for which were recorded between 1771 and 1784 [15]. During this time, Messier observed 103 objects, including star clusters and nebulae, of which 34 were galaxies. During Messier's time, and until the early twentieth century, galaxies were classified as nebulae, with some designated as "spiral nebulae" because, with improved optics, the spiral structure of these objects began to become apparent.

William Herschel, assisted by his sister Caroline, made a survey that lasted over four decades. This resulted in the discovery of far more deep sky objects than those of his predecessors, a discovery that really began to open up the universe as it truly is. Yet the cosmos in Herschel's time consisted only of the Milky Way and all that was in it, including what were thought to be mere nebulae (galaxies) or solar systems being formed. Herschel published his grand work as three separate lists in 1786, 1789, and 1802. A grand total of 2,500 objects were surveyed, of which 2,100 were galaxies.

William's son John conducted a survey of his own, between 1825 and 1833, which was mainly a re-examination of his father's original discoveries. He added 500 objects of his own finding to the list, most of which were galaxies. John extended his survey between 1834 and 1838 to cover the southern celestial hemisphere. He used his own 18-3/4-in. reflector from a site just north of Cape Town, South Africa. He observed and recorded 1,700 deep sky objects, almost all of which were new discoveries.

From 1847 to 1897 many new observers became active, multiplying the known number of deep sky objects tremendously; most of the new discoveries that were made were galaxies. In the middle part of the nineteenth century, John Herschel decided to compile a new comprehensive listing of objects resulting in the *General Catalog of Nebulae and Clusters of Stars* published in 1864. This listing contained 5,079 entries and was the prime reference work until J. L. E. Dreyer's catalog replaced this 30 years later.

In the meantime, an increase in astronomical interest in Europe and America blossomed, as larger 'scopes, better optics, and the advent of photography resulted in a still greater increase in the number of newly discovered deep sky objects. Some examples of surveys follow:

- H. L. D'Arrest published extremely accurate observations in 1867 "Siderum Nebulosorum," all of this done with an 11-in. refractor at Copenhagen.
- The Earl of Rosse in 1848 completion of a 72-in. reflector in Ireland, the largest ever built at this time, not to be surpassed until the completion of the 100-in. at Mt. Wilson. Many new, faint objects were found in the same eyepiece field of existing objects. Nearly all of these new deep sky objects were galaxies. The large aperture enabled structure to be seen in the brighter nebula-giving rise to the description 'spiral nebulae.' The survey took 30 years to complete and the results were published in the *Scientific Transactions of the Royal Dublin Society* in 1880. The obstacles that were overcome included limited observations to within one hour of the central meridian and the poor climate in Ireland.
- William Lassell, assisted by Albert Marth, on the island of Malta, used a 48-in. to discover 600 new nebulae, mostly faint galaxies, published in the *Philosophical Transactions* in 1864.

New and skilled observers extended the search to still fainter and fainter objects, e.g., Barnard, Bond, Burnham, Holden and Swift (all in America), and Bigourdan, Lohse, Schmidt, Schulz, Stephan, Tempel, and Winneck in Europe (Fig. 1.2).

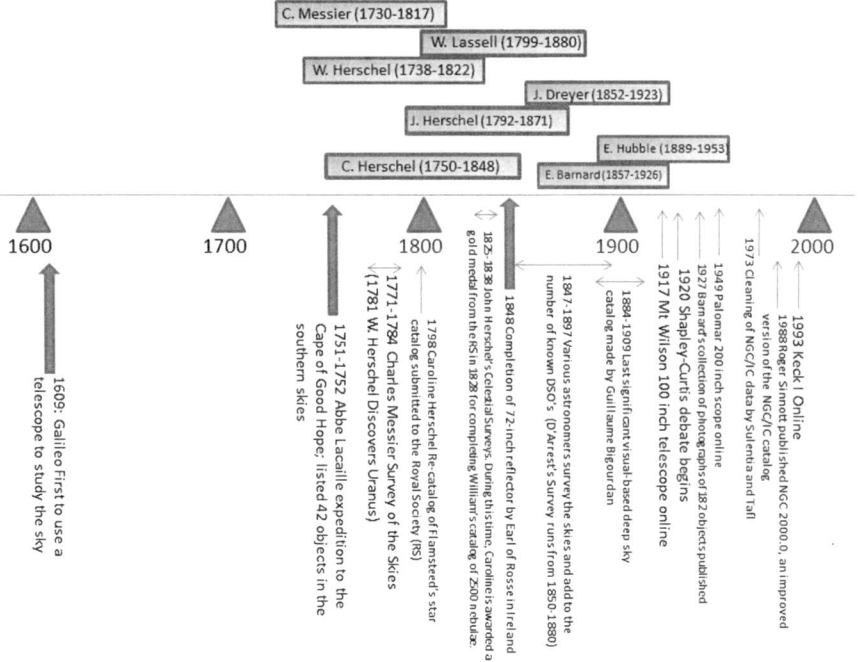

Fig. 1.2. A timeline of key events in the history of the observations of faint objects 1600–2010, as covered in the text, featuring the completion of various surveys and catalogs.

The New General Catalogue (NGC)

The catalog that is commonly used by astronomers to this day is the NGC, which contains 7,840 entries (some 85 % of the entries are galaxies, and the catalog also contains a few duplications and minor miscellaneous inaccuracies). The catalog was the result of the need to completely revise and update John Herschel's *General Catalog*. The Royal Astronomical Society promoted this project in 1886 and assigned the completion of the project to J. L. E. Dreyer, who compiled the information and published the results in 1888.

The *NGC* contains Herschel's catalog of objects, which includes 2,515 items (of these, 2,073 are galaxies). The compilation of Herschel's objects is more homogeneous, since the same individuals using the same instrument (an 18.7-in. reflector) observed most of these objects. However, the remainder of the catalog contains objects observed by a wide variety of individuals using a variety of instruments under various conditions. It is no wonder the original version of the *NGC,* along with its appendix; the *IC (Index Catalogue)* contained numerous errors, including references to non-existent objects.

This catalog remains, to this day, the universal reference standard for the brighter deep sky objects. Even after the publication of this catalog, additional discoveries were being made; as a result, Dreyer published his first supplement to the *NGC* in 1895.

The Index Catalogue (IC)

This catalog was first published by the Royal Astronomical Society and contained 1,529 objects, almost all of which were galaxies. This includes the first photographic discoveries (made by Max Wolf of Heidelburg) that have been documented. The catalog also contains visual observations made by Bigoudan, Swift, Javelle, and Burnham.

In 1908, Dreyer published the second *Index Catalogue,* listing 3,867 objects that were mostly discovered photographically. The grand total of objects included by the *IC* and the *NGC* now totaled 13,226 objects, along with positions and brief descriptions. Of these objects, about 10,000 are galaxies. The *NGC* galaxies are mostly from 13th to 15th magnitude, with the *IC* galaxies being fainter still, since most of these were discovered photographically.

The last significant visual-based deep sky objects catalog was made by Guillaume Bigourdan from 1884 to 1909. This included a survey of all the NGC objects, done with the 12-in. 'scope in Paris. Features include accurate positions and lots of valuable information useful to identify faint galaxies. Already by the start of the twentieth century, photography was surpassing visual in terms of the preferred scientific observation method of the day.

The *Index Catalogue* comes in two parts: *IC I* = Index Catalogue I and *IC II* = Index Catalogue II, but commonly you will see objects listed simply as (for example) *IC 1234*. With the exception of minor corrections, no general revisions of the *NGC/IC* catalogs happened until 1973, when Sulentic and Tifft made the attempt to clean the data with the help of the Palomar Observatory Sky Survey (POSS). However, due to time pressure, the resultant *Revised New General Catalogue (RNGC)* came out even worse than the original, in some ways, as existing corrections were ignored and new errors introduced to the *RNGC.*

Roger Sinnott in 1988 published an improved version of this catalog, in a work entitled *NGC 2000.0,* which had flaws of its own, due to time pressure again. However, the flaws were not as glaring as the *RNGC* of 15 years earlier. Today, the best source for information on these objects is a piece entitled the *Revised New General Catalogue and Index Catalogue.* This work, which now contains a total number of 14,000 entries, was assembled by an international team of amateur and professional astronomers. Each object (that is, for existing objects) has the latest data, catalog cross-references, and re-measured coordinates using the Digitized Sky Survey (DSS) to a precision of 1.2″. Other pieces of data include constellation, magnitude (B, V, V′), diameters (a, b), position angles, and Hubble classification type.

This catalog in its original form consisted of 2,712 rich clusters of galaxies and was published in 1958 by George O. Abell (1927–1983) as part of his PhD thesis at the California Institute of Technology [16]. The objects were obtained by means of visual inspections, with a 3.5× magnifying lens, of the red 103a-E plates of the Palomar Sky Survey. Continued inspection of the POSS plates in the mid-1960s revealed 86 planetary nebulae, many of which are very faint to begin with but also have a large apparent size and low surface brightness, making them a visual challenge. However, later analysis of Abell's planetary nebula catalog revealed that at least four of the objects (Abell 11, 32, 76, and 85) are not even planetary nebulae [17].

For the galactic counterpart of the survey, a cluster had to satisfy four criteria in order to be included in the Abell catalog:

Richness: A minimum population of 50 members within a magnitude range of m3 to m3 + 2 (where m3 is the magnitude of the third brightest cluster member galaxy). However, the final catalog included many clusters with fewer than 50 members. The clusters were divided into six groups, based on the richness of the clusters (again galaxies within the specified magnitude range):

- Group 0: 30–49 galaxies
- Group 1: 50–79 galaxies
- Group 2: 80–129 galaxies
- Group 3: 130–199 galaxies
- Group 4: 200–299 galaxies
- Group 5: more than 299 galaxies

Compactness: A cluster should be compact enough so that its 50 or more members lie within one "counting radius" of the cluster's center (known as the "Abell radius"). This is defined as 1.72/z arc minutes where z is the cluster's redshift.

Distance: The cluster should have a z between 0.02 and 0.2 (which corresponds to a recessional velocity of between 6,000 and 60,000 km/s or, assuming a Hubble constant of 71 km/s/Mpc, 85 Mpc and 850 Mpc). However, many of the clusters in the catalog have been found to be more distant, some as distant as 1,700 Mpc. The clusters were divided into seven "distance groups" based on the apparent magnitudes of their tenth brightest members:

- Group 1: mag 13.3–14.0
- Group 2: mag 14.1–14.8
- Group 3: mag 14.9–15.6
- Group 4: mag 15.7–16.4
- Group 5: mag 16.5–17.2
- Group 6: mag 17.3–18.0
- Group 7: mag > 18.0

Galactic Latitude: The galaxy clusters had to be sufficiently far enough above and below the plane of the Milky Way Galaxy because of the presence of high density stars (as well as interstellar extinction), but a firm parameter was not set. The galaxy cluster had to be positively identified to be included, and that included several close to or in the galactic plane that met the other criteria.

The original catalog was limited in sky coverage to declinations north of −27°, which was the original southern limit of the POSS. An additional catalog, the *Southern Survey,* was included to add 1,361 rich galaxy clusters and the IIIa-J plates of the Southern Sky Survey (SSS) were used to identify these clusters. The plates were taken with the 1.2 m Schmidt Telescope (United Kingdom) at the Siding Spring Observatory in Australia in the 1970s. This catalog was completed by Ronald P. Olowin of the University of Oklahoma and published in 1989, 6 years after Abell's death. A revised, corrected, and updated version of the original catalog was included as well as the Abell Supplement, which consisted of 1,174 additional clusters from the Southern Survey that were not rich enough or were too distant to be included in the main catalog.

The Principal Galaxies Catalogue (PGC)

This catalog consists of entries for 73,197 galaxies, which include B1950 and J2000 equatorial coordinates and cross-identifications. This was first published in 1989 by Paturel et al., and includes data such as morphology, apparent major and minor axes, apparent magnitudes, radial velocities, and position angles. It is a combination of several catalogs, including the *MCG (Morphological Catalog of Galaxies), UGC (Uppsala General Catalog),* and the *CGCG (Catalog of Galaxies and Clusters of Galaxies).* A second version of the *PGC* was published in 1996 and contained 108,792 entries.

A later version of this catalog was published in 2003 and is restricted to confirmed galaxies, about one million of them brighter than a B-magnitude of ~18. The project continued as LEDA (Lyon Extragalactic Database) and now has three million entries. Most of the data have been adopted from their source catalogs so the quality of the entry depends on the source. The *PGC* portion of this catalog gives type, magnitude, size, position angle, and cross identifications for its galaxies [18, 19].

Uppsala General Catalogue (UGC)

This catalog of galaxies, published by Peter Nilson in 1973, is essentially complete to a limiting diameter of 1.0 arc minute and/or a limiting magnitude of 14.5 [20]. It lists a total of 12,940 objects. The information for this catalog is from the blue prints (not construction plans but pictures taken through a blue filter) of the Palomar Observatory Sky Survey (POSS) and is limited to the sky north of declination −2.5°. There may be some galaxies that are smaller than 1.0 arc minute but brighter than 14.5 magnitude included in this catalog from another catalog called *Catalog of Galaxies and of Clusters of Galaxies* (CGCG, Zwicky et al. 1961–1968). A supplement to this catalog *(UGCA)* contains 444 objects of special interest south of the declination limit.

Fritz Zwicky assembled the *Catalog of Galaxies and of Clusters of Galaxies (CGCG)* between 1961 and 1968, and it contains 9,134 galaxy clusters. This and the *MCG* (1962–1974, published by Vorontsov-Velyaminov) are two classic general catalogs of galaxies based on visual inspections of the POSS data. The *CGCG* lists 29,378 galaxies as well as the above-mentioned galaxy clusters and is essentially complete down to magnitude 15 (photographic magnitude limit of 15.7) and north of −3.5° declination.

The *MCG* contains a listing of 31,917 galaxies north of −45° declination, and it is stated to be complete down to magnitude 15 (although some galaxies as faint as magnitude 20 are included). The position accuracy is only about 1–2 arc minutes, but Corwin published, in 1998, *Accurate Positions for MCG Galaxies,* which list these positions for 4,147 galaxies.

The Perek-Kohoutek (PK) Catalog of Galactic Planetary Nebulae

This catalog contains 1,510 objects that are classified as galactic planetary nebulae up to the end of 1999. This is the updated version of the *Catalog of Galactic Planetary Nebulae* [21] and includes the objects classified as planetary nebulae or possible planetary nebulae up to 1965. Six supplements bring the list to the total of 1,510 objects. Of these objects, 1,183 are planetary nebulae with a high degree of confidence, and 327 are possible planetaries (in other words, not quite the confidence needed to say with certainty that these are indeed planetary nebulae). The supplements of the original catalog contain objects called proto-planetary nebulae, virtually unknown prior to the 1980s.

The Revised and Updated Catalog of Quasi-stellar Objects (QSO)

This catalog by A. Hewitt and G. Burbidge (1993) contains information on all known quasi-stellar objects (QSO's) with measured emission redshifts, along with BL Lacertae objects, up to December 31, 1992. There are 7,315 objects, about 90 of which are QSOs (the 90 objects are BL Lac objects). Information contained in the catalog is extensive: names, positions, magnitudes, colors, emission-line redshifts, absorption, variability, polarization, and X-ray, radio, and infrared data [22].

The Sharpless Catalog (Sh)

The catalog comes in two parts, *Sh 1* (1953) and *Sh 2* (1959). The latter is a listing of 312 HII regions (emission nebulae) that is comprehensive north of (but a few are just south of) the −27° declination line. The American astronomer Stewart Sharpless published the final version of this catalog in 1959. The first version was published in 1953 with 142 objects. The objects in the catalog overlap with objects from many other catalogs, including the Messier catalog, the *NGC*, the Caldwell catalog, and the *RCW* catalog. The *RCW* catalog ('RCW' stands for Rodgers, Campbell, and Whiteoak) is a catalog of emission regions primarily covering the Southern Hemisphere, and itself consists of 182 objects (including many of the 85 items in the earlier *Gum* catalog) assembled in Australia in the 1960s [23].

The Caldwell Catalog

This catalog of 109 objects was assembled by Sir Patrick Caldwell-Moore and was meant to complement the Messier catalogue. Patrick Moore, as he is more commonly known in astronomical circles, realized that Messier's list did not include a good number of the sky's brightest deep-sky objects. These objects include the Hyades, the Double Cluster (NGC 869 and NGC 884), and NGC 253, and also bright Southern Hemisphere objects such as omega (ω) Centauri, Centaurus A, the Jewel Box, and 47 Tucanae. This catalog was compiled and published in *Sky & Telescope* magazine in December 1995.

Other Catalogs

Shortly after 1908 nebulae investigators became aware of a large number of additional objects that exist but were generally ignored in favor of brighter objects that promised more detail and scientific return. In 1918, H. B. Curtis published a paper (a part of Volume XIII of the Lick Observatory Publications) describing 762 nebulae and star clusters. As it turns out, all the nebulae were actually galaxies, and all 762 objects were already included in the *NGC* and *IC* catalogs. Curtis acknowledged additional galaxy images visible on his plates.

Not much attention was given to these new, anonymous galaxies until after the completion of the initial Palomar Sky Survey in 1956. This survey revealed the immense number of uncataloged galaxies in the universe. As a result, catalogs are updated to include various types of deep sky objects. One particular image has three objects that were solely galaxies (pp. 1–2 of Volume 6), each with first-time listings of new or 'anonymous' galaxies. In fact, more recently, the Hubble Deep Field images (one example is shown in Fig. 1.3) from the north and the south turned up even more galaxies in huge numbers, greatly increasing the estimate of the number of galaxies in our known universe.

Other examples of catalogs include those of special types of galaxies such as the *David Dunlop Observatory Catalog* of 243 dwarf galaxies, *Catalog of Dwarf*

Fig. 1.3. The Hubble Ultra Deep Field showing lots of "beyond-faint" galaxies (Image courtesy of NASA: Hubblesite.org).

Galaxies by Karachentseva (which is somewhat larger than the *DDO* with 260 objects), a collection of 104 spherical dwarf galaxies by Mailyan, and Tully's *Nearby Galaxy Catalogue* containing 2,367 such objects great and small. Edge on or superthin galaxies are featured in the *Revised Flat Galaxy Catalog* (there are 4,444 entries and this supersedes the *Flat Galaxy Catalog*) by Karachentsev. Then there's *Arp's Atlas of Peculiar Galaxies* (all 338 of them), the *Atlas of Interacting Galaxies* (852 entries in two parts) by Vorontsov-Velayminov. And who can forget *Zwicky's Catalog (Zw) of Selected Compact Galaxies and of Post-Eruptive Galaxies* (1–8 Zw; approximately 3,000 objects).

For folks "down under" we have the *Atlas of Southern Peculiar Galaxies and Associations* by Arp and Madore, featuring 6,445 entries. Then there is the *Atlas of Polar Ring Galaxies* (*PRC*, 157 entries), which provides lots of examples of this fascinating type of galaxy. In addition, catalogs of pairs, groups and clusters of galaxies are recorded; some examples include the *Catalog of Isolated Pairs of Galaxies* (*KCPG*, 603 entries), *Catalog of Isolated Triplets of Galaxies* (*KTCG*, 84 entries), and the *Catalog of Isolated Galaxies* (*KARA*, 1,106 galaxies). Each of these three were published by the Karachentsevs (the former two followed the latter).

"Exotic" objects, such as those in BL Lacertae, active galactic nuclei (AGN), and quasars are also cataloged. The *Catalog of Quasars and Active Galactic Nuclei* by Veron-Cetty and Veron contains information on each type of object, and its 13th edition was released in 2011. No information about absorption lines or X-ray properties of these objects is given; absolute magnitudes (assuming a Hubble constant of 71 km/s/MPc) are provided along with a deceleration parameter q0 = 0. This catalog contains a total of 168,941 objects of which 133,336 are quasars, 1,374 are BL Lac objects, and 34,231 are AGN (including 15,627 Seyfert galaxies).

Still more catalogs include the *Revised and Updated Catalog of Quasi-stellar Objects* by Hewitt and Burbidge (1993), which contains 7,315 objects; AGN catalog has "faint blue stars" and includes these lists: Humason and Zwicky (48 entries), Tonantzintla (419 objects), Usher (this lists 2,363 objects), Palomar-Haro-Luyten (with a total of 8,725 entries), Palomar-Berger (has even more entries, 9,495), and the Luyten Blue Star catalog (11,444 "blue stars" included). Searches for special objects or galaxies that display a strong ultraviolet (UV) flux and/or emission lines resulted in many new lists being produced, including the Markarian (1,515 objects, also known as the First Biurakan Survey), the Arakelian (591 objects), the Kazarian (466 objects), the Haro (44 objects), and the Palomar-Green (1,874 entries).

Conclusions and Reflections

The above is a sampling of the many catalogs that are out there that attempt to document what is out there in the nighttime sky. These catalogs are either general (including the range of types of deep sky objects) or more specific (focusing on a specific type of star or object), but their purpose is to document what we know that exists in the galaxy and beyond. The earliest known star catalog was one consisting of 36 stars compiled by the ancient Sumerians between 1555 and 1531 B.C. The Greek astronomer Hipparchus assembled his catalog of some 850 stars in 129 B.C., and he ranked them in terms of brightness, which (after a few modifications) led to the current stellar magnitude system. A number of other star catalogs have been made, before and after Hipparchus, until the age of the telescope.

As impressive as the numbers of objects may seem to be, all of their numbers combined are but a drop in the ocean or a grain of sand from all the world's beaches (or whatever tiny piece of a vastly big example you want to use). By one estimate, there are 200 billion galaxies in the visible universe, and each galaxy has an average of 100 billion stars. Each galaxy, depending on its contents, has its own collection of billions of emission nebulae, planetary nebulae, star clusters, supernova remnants, etc. On the one hand, who knows what sorts of spectacular objects lay unseen in some inaccessible alien sky; on the other hand the laws of physics are the same throughout the universe. As a result, the possible ranges of what we have access to in the night sky, even with the largest amateur 'scope, looking for the faintest objects with electronic detectors, is but an infinitesimally small sample of what is really out there (refer to the figure below). Mostly unseen is the vast majority of the universe, some of which is only beginning to be revealed with modern telescopes, technology, and techniques in wavelengths unseen to the human eye. Billions of planets unseen could lurk in our home galaxy alone.

Chapter 2

The Astronomers Behind the Historical Surveys

Introduction

What follows is a brief historical narrative of some of the more important figures in the history of deep sky astronomy during the age of the telescope. Far from being a complete listing, the following does highlight some of the important figures involved in expanding our view of the universe. Once the telescope was invented and accepted as a useful scientific instrument, along with improvements in the optics of both reflectors and refractors, it became possible for various astronomers to carry out systematic surveys of the nighttime sky, cataloging the various objects they encountered along the way.

Charles Messier

We begin with one of the most famous astro-surveyors, Charles Messier, of Messier object fame [25, 26]. Messier (Fig. 2.1) was born June 26, 1730, in Badonviller, Lorraine, France. He was the 10th of 12 children in his family, born to Nicolas Messier (1682–1741) and Francoise Grandblaise (d. 1765).

Messier grew up rather wealthy but at the young age of 11, he lost his father, who died unexpectedly in the year 1741. Death was no stranger to the Messier family, as six of Charles' siblings had died at young ages by 1729. Charles had an interest in astronomy as a boy and developed the skills needed in this arena; this interest was stimulated by the appearance of the Great Comet of 1744, the famous six-tailed comet of De Chéseaux. Messier also witnessed an annular solar eclipse from his hometown on July 25, 1748.

Charles Messier acquired the ability to discern fine details in administrative work, and the ability to make precise measurements, and these were further perfected when he entered the workforce, being employed by Joseph Nicolas de l'Isle (1688–1768), the astronomer of the Navy, in 1751. The work duties that Charles was given further prepared him for observational astronomy. These duties included various handwriting projects and copying (by hand) a large map of China. After familiarizing himself with the observatory of de l'Isle (to catalog its contents and instrumentation), Messier made his first documented astronomical observation,

B. Cudnik, *Faint Objects and How to Observe Them*, Astronomers' Observing Guides, DOI 10.1007/978-1-4419-6757-2_2, © Springer Science+Business Media New York 2013

Fig. 2.1. Charles Messier (Courtesy Wikimedia commons [24]).

the transit of Mercury across the solar disk on May 6, 1753. This event was one example of how exact positions are vital for scientific quality of all observations, something that Messier quickly learned to appreciate and later apply to the observations that contributed to the success of his catalog.

It is fascinating to note that this observatory that Messier became familiar with already had an interesting history by the time Messier began working there. The observatory was "…established in 1748, after Delisle's return from Russia a year before, on a tower of Hôtel de Clugny (later Hôtel de Cluny), a residence constructed in 1480, on the ruins of Roman thermes from the fourth century, as the temporary Paris residence for the abbots of the Clugny order and their guests. In the eighteenth century, it was rented to the administration of the Royal Navy [26]."

Another astronomer, Edmund Halley, predicted that the comet of 1682 would return again in 1757 or 1758, so Charles Messier began to search (unsuccessfully) for this comet, using charts that were incorrectly been prepared by de l'Isle. The search for this returning comet began in 1757, the same year he observed what would later be known as M32 and (very likely) the "Great Nebula" M31 (also known as the Andromeda Galaxy). Messier did find another comet, on August 14, but this one had been discovered 3 months prior (this was comet De la Nux, C/1758 K1, discovered on May 26, 1758). Two weeks after his comet find, Messier found what would later be known as M1, the Crab Nebula, which appears as a comet-like patch of gray haze in Taurus. The search for Halley's Comet continued over a 2-year period of 1757–1758, and this began the new astronomical discipline of comet hunting, making Messier the first comet hunter.

Charles made the Crab Nebula the first "official" entry of his new catalog. At this point, he had swept up (with his telescope) more pseudo-comets (nebulae), so he began to record the positions of these annoying objects to prevent him from mistaking them for comets during subsequent hunts. Messier continued to find more and more "non-comets" and even a few real comets. The "non-comet" objects included M2 (previously discovered by Jean-Dominique Maraldi) in 1759, Comet 1762 Klinkenberg (which he monitored from May to July 1762), and on September 28, 1763, he finally discovered his own comet, Comet Messier (1763). Messier discovered a third nebula-like object that turned out to be a new discovery (M1 and M2 were both known to astronomers prior to Messier's observations of these) that motivated him to get serious about surveying the skies and cataloging these objects.

In 1759, Charles Messier was appointed chief astronomer of the Marine Observatory. Almost 5 years later, Charles Messier found the second comet to be named Messier on January 3, 1764. During a 7-month period of comet searches this year, Messier added 38 new objects to his list, including M13 (the great Hercules globular cluster) and the Swan Nebula (M17) in Sagittarius. In all, Charles Messier (nicknamed "Comet Ferret" by King Louis XV) was the discoverer of 15 comets, and a co-discoverer of 7 additional comets. Besides these nebulous objects that now bear his name, Messier observed other astronomical events and attractions, such as the Venus transit of June 6, 1761, and the rings of Saturn. The following lists the comets that Messier discovered over the course of his astronomical career: C/1760 B1 (Messier), C/1763 S1 (Messier), C/1764 A1 (Messier), C/1766 E1 (Messier), C/1769 P1 (Messier), D/1770, L1 (Lexell), C/1771 G1 (Messier, the Great Comet of 1771), C/1773 T1 (Messier), C/1780 U2 (Messier), C/1788 W1 (Messier), C/1793 S2 (Messier), C/1798 G1 (Messier), and C/1785 A1 (Messier-Mechain).

Messier discovered 19 objects in 1765, a year after becoming a Fellow in the Royal Society and 2 years after being elected being elected to the French Academy of Sciences. Four years later, in 1769, Messier was elected as a foreign member of the Royal Swedish Academy of Sciences. By 1765 several astronomers had already compiled catalogs that Messier had drawn from to assemble his own catalog. These catalogs included the following: Edmond Halley made a list of six objects; William Derham kept a catalog of objects he had extracted from Hevelius' star catalog, *Prodomus Astronomiae*; and Nicolas Lacaille's Catalog of Southern "Nebulae" of 1755 (Messier also used lists of objects from Maraldi and Le Gentil).

Charles Messier continued his personal and professional achievements in the 1770s. On June 30, 1770, he was elected to the French Academy of Science. Almost 5 months later to the day, November 26, 1770, Charles Messier (now age 40) married Marie-Françoise de Vermauchampt (age 37) after knowing each other for about 15 years prior; the following year, Messier was appointed astronomer of the Royal Navy. Messier became a father with the birth of Antoine-Charles, on March 15, 1772. Sadly, both his wife and son died within 11 days of the boy's birth.

The first version of Messier's own catalog was published in 1774 and included 45 entries, M1 to M45. The tail end of this early version of the catalog is graced by the (already then) well-known objects (for each of these, he determined their celestial positions) M42 (the Great Orion Nebula), M43, M44 (the Beehive cluster), and M45 (the Pleiades). Of these 45 objects, 17 were Charles's own personal discoveries (the rest being objects contained in the catalogs mentioned above, discoveries made by other astronomers). Charles looked for several non-existent nebulae from older catalogs and confirmed their non-existence, and is probably at least a contributing reason for the entry of M40 that is not a nebula, galaxy, or star cluster but a double star in Ursa Major.

Only weeks after the tragedy of losing his wife and son, Charles Messier was back at his surveying work, adding object number 50 to his list on April 5, 1772. This is the open cluster later known as M50. By May 1780, at the time that Messier became a member of the Literary Society of Upsala, Sweden, he had discovered his 68th object, completing the list for the "second edition" of his catalog.

Charles Messier's catalog was almost complete in 1782, with the addition of his 107th entry; at this point in history another historic figure, an astronomer by the name of William Herschel (with help from his sister Caroline) began his own survey for deep sky objects. Messier served to motivate Herschel to begin this work that would later dwarf Messier's list in terms of numbers of entries and sheer ambition. While Messier was making his last discoveries (the last deep-sky object discovered by him was made in 1798; his last comet was found in 1801), Herschel was making his first discoveries (the planet Uranus in 1781 and his first, very own deep-sky object in 1782, now known as the Saturn Nebula, or NGC 7009).

Messier was able to view the then newly discovered asteroids 1 Ceres and 2 Pallas, and continue to discover comets (the last discovery was in July 1801). Messier was presented the Cross of the Legion of Honor by Napoleon Bonaparte in 1806 for dedicating (in a memoir) the Great Comet of 1769 to Napoleon, who was born that year. At this point it is interesting to note that Messier was probably the last true scientist to claim that comets carried an atmospheric and/or astrological significance. As Charles Messier reached the last decade of his life, he observed less until a debilitating stroke in 1815 left him partially paralyzed. He did make a partial recovery and was able to attend one more meeting of the academy of which he served as a member for 37 years up to this point. On April 12, 1817, Charles Messier passed away at age 86 at his home in Paris.

Today, Messier's catalog has a total of 110 objects, 7 of which were added in the twentieth century. The last entry, M110, was added in 1967. Being among the brightest and easiest to find objects in the nighttime sky, these are favorite observing targets for amateur astronomers. The "Messier Marathon" is done each March, when all but one or two Messier objects are accessible to amateurs who are skillful enough and have planned well enough to successfully observe the ones in evening and morning twilight. The goal is to locate, identify, and observe as many of the 110 objects in a single night. The Messier catalog includes, according to many amateur astronomers, some of the best deep sky objects observable in the northern hemisphere [26, 27].

Frederick William Herschel

The English astronomer by the name of Frederick William Herschel (Fig. 2.2) was born November 15, 1738, in Hannover, Germany [29]. His father Issak Herschel was a musician with the Hanoverian guard, and William followed in his footsteps as a musician, playing the oboe. Due to religious struggles, this family of Protestants moved from Morovia to Saxony in the early seventeenth century. William's early education was very limited due to the warlike conditions of his country at this time. He became a self-taught mathematician and enjoyed considerable scientific

Fig. 2.2. Sir William Herschel [28].

success. He also became a skillful musician, excelling in both music theory and practice.

Being an accomplished musician, William Herschel composed his own music and gave concerts; however, his primary interest, his "first love," soon became astronomy. He educated himself to become an astronomer through the writings of Ferguson, Keill, and eventually to those of Lalande. His mathematical training helped prepare him for his astronomical pursuits. In 1752, at age 14, he joined the band of the Hanoverian guard, who, along with William, visited England in 1755. Due to the Seven Years' War compromising health and quality of life, William's parents moved him to England on July 26, 1757. He did odd music-related jobs in several towns in northern England during his first years of living there, but things began to look up for William in 1766 when he was appointed an organist at the Octagon chapel of Bath, a wealthy city at the time. He ascended the ranks of musicians over the next 5 years, becoming the director of all the chief public music entertainment at Bath. Having settled into a lucrative career, he returned to Hanover briefly to bring his sister Caroline Herschel to Bath, where she arrived in August 1772 at age 23. From then on, Caroline would serve as William's assistant in his many and varied pursuits.

Even as an accomplished musician in a position of authority, William Herschel continued to study texts of music and astronomy (Smith's *Harmonics* and Feguson's *Astronomy* were two of his favorites). He continued to educate himself to become an astronomer by reading the works of Ferguson, Keill, and later Lalande. During Herschel's time, telescopes were of poor optical quality by today's standards; they were also very rare and very expensive. As a result, he started with a small 2-in. (5-cm) Gregorian reflecting telescope that he rented, but soon sought larger instruments. He purchased a small lens with a focal length of about 18 ft (and gave it to his sister Caroline to make a tube for it, resulting in it being usable as a telescope). This telescope, however, produced poor images of Jupiter, Saturn, and the Moon. After looking unsuccessfully for a much larger, affordable reflector, Herschel decided to build his own large telescope. This was not accomplished at once, as he needed to work his way from polishing small mirrors to, by 1774, completing a Newtonian telescope with a 6-ft (1.8-m) focal length that gave him satisfactory views of the heavens.

Starting that same year, 1774, Herschel spent every available hour of the night systematically exploring the skies and meticulously recording his observations, with the help of his sister Caroline. He envisioned surveying the entire sky and finding the overall structure of the universe through his observations. During the "off season," when tourists were no longer visiting Bath (he had been kept busy performing for them as a musician), he spent the daylight hours, with the help of his brother and sister, grinding and perfecting mirrors. His observations occupied every hour of the night until his discovery of the planet Uranus on March 13, 1781 (this was made with a 7-ft-/2.1-m-long, 6.5-in./15.5-cm wide reflecting telescope). He thought this was a comet at first, but after months of observations and calculations of the object's orbit, he concluded it was a planet in a near-circular orbit well beyond the orbit of Saturn. He also observed the object with his larger, 20-ft (6.1-m) long 'scope, and in so doing discovered two of the moons of Uranus. He eventually was able to construct his biggest telescope, a 48-in. (1.23-m) reflector, but he made the most of his recorded astronomical observations with his 20-ft (6.1-m) long reflecting telescope (Fig. 2.3).

Not only did Herschel make systematic observations of objects by night, and mirrors by day, he was also able to find time to write professional papers for the Royal Society. In May 1780 he published his first two papers with the Royal Society, one including his observations on the variable star Mira and the other concerned with the lunar mountains. Herschel studied variable stars and used them to try and find out what is going on with the Sun. Knowing of the sunspot occurrence on the Sun and the fact that the Sun rotated, Herschel asked himself a series of questions concerning variable stars and used these questions to guide his observations (observations motivated by self-guided inquiry). He published a series of papers beginning in 1781 on the rotations of the planets and their satellites (that is, those that were known at the time). He published seven more papers over the next 18 years, which included measurements of the rotation period of Mars, the discovery and observations of the polar caps on Mars, and the time it takes for natural satellites to complete one orbit around their parent planets. He used guided inquiry all along, and was very systematic in his investigations.

The telescopes that Herschel used did not have clock drives like modern instruments, so he observed in the following manner: He set the 'scope at a certain elevation, pointed toward the meridian, and watched what drifted through over the course of the night. Being on a ladder, he called his observations out to his sister Caroline, who recorded the time and the information. He was able to cover a thin strip of sky, and then on subsequent nights, he would change the elevation and repeat the exercise. Eventually he covered the entire celestial sphere that is visible from Great Britain.

Over the course of his astronomical career, Sir William Herschel discovered many faint objects called nebulae and systematically cataloged these objects. He made many astronomical discoveries: two of Saturn's moons, nearly 1,000 double stars (as well as the finding that these orbit around a common center of gravity

Fig. 2.3. Herschel's huge telescope with which he did his surveys.

after repeated measurements over an 11-year period), the movement of the Solar System through space, and the confirmation of the gaseous nature of the Sun (which he arrived at with the help of his observations of sunspots). He attempted to describe the overall structure of the universe, starting with the intuitive idea that the brighter stars are closer to the Solar System, and the fainter ones are further away. Herschel understood the concept behind trigonometric parallax, the most fundamental method astronomers have to measure the distances to the stars, but was unable to actually make use of this method.

King George III, in 1782, invited Herschel to Windsor to hire him as his private astronomer. As a result, Herschel was able to work full time in science and moved from Bath to Datchet and ultimately to Slough, where he was able to easily keep in contact with the king. He was able to continue his astronomical pursuits. In 1783, Herschel first considered the "Motion of the Solar System in Space," a paper that considered the idea of a moving Solar System among the stars of the Milky Way. He revisited this idea in 1805 and published details on the construction of the cosmos and ultimately arrived at the idea that the Sun is a star near the galactic plane and that all the visible stars occur in clusters that are scattered along the galactic plane.

Herschel married Mary Pitt, the widow of a wealthy London merchant, John Pitt, on May 8, 1788. They had one son, John Herschel, who would later carry on his father's work of surveying the heavens and cataloging its contents. John took his father's telescope and support structure to South Africa and used them to survey the southern skies that are not accessible to the UK. In 1864, John published a catalog of 5,097 objects, 4,630 of which were discovered by John and his father. This catalog of nebulae came to be known as *The General Catalog of Nebulae.*

Sir William Herschel died at Slough on August 25, 1822, at the age of 83 and was buried under the towers of St. Laurence's Church in Upton, Great Britain [30].

Caroline Lucretia Herschel

Caroline Herschel (Fig. 2.4) was born on March 16, 1750, in Hannover, Germany, and was one of ten children, a younger sister of her brother William [31, 32]. She helped out in the management of her mother's household until 1772, when she left to join her brother William in his astronomical pursuits. She showed talent for music and mathematics and eventually began her own surveys of the skies, discovering eight comets of her own between 1786 and 1797.

Her health and growth were impacted by a bout with typhus at age 10, remaining short (no more than 4 ft 3 in. or 1.3 m tall) and frail all her life. She was able to perform as a soprano and gave concerts with her brother William in the 1770s, shortly before the two of them began to work together in astronomical surveys. Carline's role in the surveys was to record his observations at night and finalizing them during the daytime. She also planned the observing schedule for each night.

During the times of William's absence on business trips, Caroline began to spend more time on the telescope herself, which led to her increasing interest in observational astronomy and the comet discoveries mentioned above. She found her first deep space object in February 1783, an open star cluster known today as NGC 2360. Her second object was found in August 1783—a galaxy that is referred to today as NGC 205 (also known as M110), an object that Charles Messier took note of in 1773 but did not record in his list of objects. Carline found a number of additional objects in the fall of 1783: NGC 253 (the Sculptor Galaxy), NGC 381, NGC 659, NGC 2548 (M48), NGC 6633, and NGC 7789.

Caroline found NGC 225 in 1784 and NGC 7380 in 1787, and her first comet was found in August 1786, while her brother was in Germany. Caroline became the first woman to be officially recognized for a scientific position as assistant to William. In addition to discovering the 8 comets over 11 years, Caroline began re-cataloging Flamsteed's star catalog, submitting it to the Royal Society in 1798, along with an additional 560 stars that Flamsteed did not include with this catalog. After her brother's death in 1822, she completed his catalog of 2,500 nebulae and received the Gold Medal of the Royal Astronomical Society on February 8, 1828.

Fig. 2.4. Caroline Herschel [31].

Caroline received other awards and recognitions later in life, which included being the first woman to receive honorary membership into Britain's Royal Society in 1835, being elected into the Royal Irish Academy in 1838 and awarded the Gold Medal for Science by the king of Prussia in the year 1846. Caroline returned to Hannover and died on January 9, 1848. She left a truly stellar legacy as one of the first women to be recognized for many achievements.

John Herschel

Sir John Frederick William Herschel (Fig. 2.5) was born on March 7, 1792, in Slough, Buckinghamshire, England, the only son of William Herschel and Mary Pitt. Caroline Herschel helped tremendously in John's upbringing [34]. Although schooling presented challenges to John, including being on the receiving end of bullying at Eton College at the age of 8, he completed his studies at Dr. Gretton's School in Hitcham and Clewer and with the help of a private mathematics tutor, was prepared for university. John enrolled at St. John's College in Cambridge in 1809, where (as an undergraduate) he met two famous mathematicians named Peacock and Babbage. In 1812, these three undergraduates founded the Analytical Society, whose chief purpose was to introduce a specific mathematical theory, known as the Continental Method of mathematical analysis, into English universities. This society only lasted a few years, but John Herschel's involvement in mathematics continued for many.

Fig. 2.5. John Herschel (Courtesy www.scientific-web.com [33]).

Herschel graduated from college in 1813 and pursued his interest in law. He began training in the legal professions in 1813 after graduating from college, but he gave this up after 18 months to return more fully to mathematics. That same year he was elected as a fellow of the Royal Society of London for his mathematics paper entitled "On a remarkable application of Cotes's theorem," which was published in the Transactions of the Royal Society. Herschel studied algebra and published papers on trigonometric series. He also published a two-volume book in mathematics in 1820, among his many other mathematical works. However by this time (and even before) he began to show interest in other subjects, which began to draw him away from mathematics.

John Herschel demonstrated a variety of abilities, each of which he showed himself to be capable at, but he had not made a significant advance in any of these. Because of the many areas he was working in, he usually moved from one to another before making a significant advance in the first one. Had he continued fully in mathematics, he may have become one of the most notable mathematicians of the nineteenth century. His interests focused on astronomy during the summer of 1816, a move that was most likely the result of the ailing health of his 78-year-old father who had accomplished so much in observational astronomy, yet there was much that remained to do, and there was nobody else who was willing to continue this work. On October 10, 1816, he wrote the following to Babbage:

> I shall go to Cambridge on Monday where I mean to stay just enough time to pay my bills, pack up my books and bid a long—perhaps a last farewell to the University. …
> I am going under my father's directions, to take up the series of his observations where he has left them (for he has now pretty well given over regular observing) and continuing his scrutiny of the heavens with powerful telescopes …

From this point on, John Herschel began his work in astronomy while studying other topics. Herschel was involved significantly with the founding of the Astronomical Society in 1820, where he was elected vice-president at the second meeting of the society. He was awarded the Copley Medal of the Royal Society of London for his mathematical analysis work.

Among his many other activities was his work with photography, specifically the chemical processes involved with this, which was foundational in the development of photography. Herschel had so many talents that spanned so many subjects that he was not able to make those last breakthroughs to be considered as the inventor of photography. He conducted and published this work in 1819, and published more papers on the subject in 1839, 1840, and 1842. By the way, the first permanent photograph was taken in France by Joseph Nicéphore Niépce in 1826 after much experimentation, but he did not have this or any of his inventions officially recognized by the time of his death in 1833. John Herschel also had great skill as an artist, so his need to invent photography was less urgent than for most others.

John Herschel also loved the outdoors and traveled with his good friend Babbage from college to Italy and Switzerland in 1821; they also visited other notable scientists of the time, including Arago, Laplace, Biote, Fraunhofer, and Talbot. Both Babbage and Herschel loved to climb mountains, but Herschel's exploits of the great outdoors was as much a learning experience as it was recreational.

In 1822, the year of his father's death, John Herschel published his first paper in astronomy, which was about a new method to calculate lunar eclipses. This was followed in 1824 by his first major publication, a catalog of double stars, which earned him honors from the Royal Society. In addition, the Paris Academy awarded

John their Lalande Prize in 1825 and the Astronomical Society awarded him their Gold Medal in 1826. In the year he won the Gold Medal he published an important paper on the topic of stellar parallax, but he himself was not successful in determining the parallax of any star (it was Bessel who would be first to demonstrate this, and he published the results of his study in 1840). He continued his father's work on double stars, to include developing methods to calculate orbits of these stars; he was awarded the Royal Medal of the Royal Society in 1833.

John Herschel published his *Discourse on Natural Philosophy,* one of his most important works in 1830; Michael Faraday describes this in a letter to Herschel:

> … when your work on the study of Natural Philosophy came out, I read it, as others did with delight. I took it as a school book for philosophers, and feel that it has made me a better reasoner and even experimenter, and has altogether heightened my character, and made me, if I may be permitted to say so, a better philosopher.

John Herschel was heavily involved with the Royal Society and that involvement had significant influence on his career. He served as secretary of the society from 1824 to 1827 (when he resigned), and in 1831 was nominated for president. He lost by a narrow margin owing to an ongoing battle between two schools of thought in the society, the traditionalists and the reformers. John was a reformer, but the traditionalists were slightly greater in number, resulting in Herschel's defeat in 1831. He had been elected president of the Astronomical Society in 1827 and was knighted in 1831, so he had his share of honor. But the 1831 Royal Society event likely motivated a decision by Herschel to go to the Cape of Good Hope in South Africa.

After the death of his mother, Mary Herschel in 1832, and the completion of the Royal Observatory at the Cape of Good Hope 4 years earlier, John Herschel set sail for South Africa in 1833. He decided to set up his own 20-ft (6.15-m) refracting telescope in South Africa to survey the skies for objects not accessible from the Northern Hemisphere. His family accompanied him in this voyage to South Africa, arriving in January 1834, since he married his wife Margaret Brodie Stewart 4 years prior to the voyage.

He made many important discoveries, including his observations of Comet Halley in 1835, where he recognized the major forces perturbing the comet, repelling it from the Sun. This force could be considered the discovery of the solar wind, which later was found to be the reason for that repulsive force. Herschel also discovered that gas was evaporating from the comet. That same year, on September 9, Herschel, building upon significant developments by William Henry Fox Talbot, Joseph Niépce, and Louis Daguerre (the person for whom the early photos called "Daguerrotypes" are named) exposed the first glass plate. This was but one of many contributions that John Herschel made to the technologies and techniques of photography.

John Herschel returned to England in 1838 but did not have much time for his many astronomical observations. He was so busy with other events (partly non-scientific) in his life to the extent that he was not able to complete the task of reducing his South Africa observations until 1847. For his efforts in the form of another important publication, Herschel received his second Copley Medal from the Royal Society.

John Herschel heard of Daguerre's work on photography in January 1839, and within a few days he had learned how to take photographs (back then, it was more involved than simply "point and shoot"). He was able to learn quickly because of work he did in 1819 on the chemical processes related to photography. Herschel published more papers on the subject in 1839, 1840, and 1842; he had done much work with the development of photography, but like other endeavors in his life, he did not take the further steps required to being recognized as the inventor of photography. Several words related to photography, including the word itself (photo = light, graphy = drawing), "negative," "positive," and "snapshot" were first used in his paper entitled "Note on the art of Photography, or The Application of the Chemical Rays of Light to the Purpose of Pictorial Representation," presented to the Royal Society on March 14, 1839 [35].

In 1842, John Herschel accepted a position as rector of Marischal College in Aberdeen, England; he also served as president of the British Association at Cambridge in 1845. In addition he accepted a position in 1850 that was "out of his league," being a scientist by nature and not a businessman. This was head of the Mint, and this happened during hard times, when major changes were afoot, but the manner in which these changes would be carried out was not yet settled upon. Herschel had difficulties dealing with most aspects of the job (the staff and Treasury, specifically), and this prevented him from pursuing his scientific interests. After a few years of difficulty and separation from his family, he resigned from this post, but suffered declining health as a result. He "retired" at age 63, but not from scientific work (and it is interesting to note that he never held any paid scientific position).

John Herschel died on May 11, 1871, in Hawkhurst, Kent, England.

With his many astronomical achievements, and his work in England, John Herschel made no major breakthroughs in astronomy that he is remembered for, but he was greatly admired by his now-famous colleagues (Biot, Babbage, among others). In fact, Biot considered John Herschel to be the natural successor to Laplace when Laplace died in 1827. Biot wrote in the obituary: "In John Frederick William Herschel British science has sustained a loss greater than any which it has suffered since the death of Newton, and one not likely to be replaced…" And Tait wrote of John: "Every day of Herschel's long and happy life added its share to his scientific services."

The obituary then goes on to say that of even higher value than his research was the influence of his teaching and example in "wakening the public to the power and beauty of science, and stimulating and guiding its pursuit."

John Lewis Emil Dreyer

John Lewis Emil Dryer (Fig. 2.6) was born on February 13, 1852, in Copenhagen [36]. His father was John Christopher, Danish Minister for War and the Navy.

From the days of his early schooling, John Dryer showed an unusual ability in the areas of history, mathematics, and physics. These subjects provided the foundation needed for his later work and would eventually form the background to his work later in life. He accepted the position of assistant to Lord Rosse (the son and successor of the Lord Rosse who built the "Leviathan" telescope located at Parsonstown) in 1874 at the age of 22, at Birr, John had access to the Leviathan, a 72-in. (1.8-m) 'scope that was, at that time, the largest in the world. With this 'scope, he started a comprehensive survey of star clusters, galaxies, and nebulae. He served as observing assistant at Dunsink Observatory (1878–1882) prior to becoming

Fig. 2.6 John Lewis Emil Dreyer (Courtesy of Armagh Observatory).

director of Armagh Observatory in 1882, a position he kept until 1916. He married Katherine Tuthill of Kilmore County, Limerick, in 1875, and the couple had three sons and one daughter.

During his time at Armagh Observatory, the facility struggled financially to the point of not being able to replace its aging instruments. Dryer, however, was able to obtain a new 10-in. (25.4-cm) refractor but was unable to make full use of this because of a lack of funding for an assistant. He then focused on the completion of the survey he initiated earlier that would become the *Second Armagh Catalog of Stars and Nebula,* his most important contribution to astronomy. This work would also become a key component to the *New General Catalogue of Nebulae and Clusters of Stars (NGC).* In this catalog are listed 7,840 objects and would later be supplemented by two *Index Catalogues* (1895 and 1908), which lists 5,386 more objects.

In addition to his astronomical observations, Dreyer was interested in the history of astronomy, especially the work of Tycho Brahe. Dreyer published, in 1890, a work about the life and work of Tycho, which became the standard biography in English. He later completed his most notable work a 15-volume compilation of the complete works of Tycho (written in Latin). He also published *The History of the Planetary System from Thales to Kepler,* another classic of historical astronomy. Dryer died 10 years later, in Oxford, on September 14, 1926. His distinctions included: Gold Medal, Copenhagen (1874), Gold Medal—Royal Astronomical Society (1916), President R.A.S. (1923–1925), D.Sc. Belfast, Hon. M. A. Oxford.

Edward Emerson Barnard

Edward Emerson Barnard (Fig. 2.7) was born on December 16, 1857, in Nashville, Tennessee, to Reuben and Elizabeth Jane Haywood Barnard [37]. His father died before Edward was born, and Edward was raised in poverty during the Civil War. Barnard had one brother and received little in the way of formal schooling. He became interested in photography and became a photographer's apprentice at the age of nine. His interest in astronomy developed in his teen years. He bought a 5-in. (12.7-cm) refracting telescope in 1876 and discovered his first comet (but he failed to announce this discovery) in 1881. He discovered a second comet that same year and a third in 1882.

While working in the photography studio, Edward Barnard married Rhoda Calvert (born in England) on January 27, 1881, and had a house built for himself and his wife with money earned for discovering five comets (at $200 per comet). Eventually, Barnard was able to obtain a fellowship to Vanderbilt University with the help of Nashville-area amateur astronomers. Unfortunately Barnard never completed his degree, but he did receive an honorary degree (the only such degree Vanderbilt has ever been awarded). He took his first professional post at the Lick Observatory in 1887, where he discovered Amalthea (Jupiter's fifth known natural satellite and the last moon to be discovered by visual means rather than examining images) and came up with wide-field photographic techniques with which to survey the heavens.

Barnard was the first to discover strong evidence for ring "spokes" in an occultation of the moon Iapetus by Saturn's rings as seen from Earth. This was manifest in the form of a shadow passing over Iapetus as it moved through the space

Fig. 2.7. E. E. Barnard.

between the innermost rings and the planet itself. The *Voyager 1* spacecraft confirmed the existence of spokes 92 years later.

Between the years 1881 and 1892, E. E. Barnard discovered 14 unique comets; 3 of these are periodic comets. He co-discovered two other comets. The comet discoveries that Barnard made included D/1892 T1 (Barnard 3), the first comet to be discovered by photography (this was recovered in late 2008 as 206P/Barnard-Boattini). Barnard's discoveries were:

C/1881 S1

C/1882 R2

D/1884 O1 (Barnard 1)

C/1885 N1

C/1885 X2

C/1886 T1 Barnard-Hartwig

C/1887 B3

C/1887 D1

C/1887 J1

C/1888 U1

C/1888 R1

C/1889 G1

177P/Barnard (P/1889 M1, P/2006 M3, Barnard 2)

C/1891 F1 Barnard-Denning

C/1891 T1

D/1892 T1 (Barnard 3)—First comet to be discovered by photography; recovered in late 2008 as 206P/Barnard-Boattini

Barnard's mother Elizabeth Jane Haywood Barnard passed away during this time (in 1884).

In 1892, Barnard became the first to realize that gaseous emissions come from novae (this was done spectroscopy, not visually), which showed these events to be stellar explosions. Three years later he joined the University of Chicago as professor of astronomy, where he was able to use the 40-in. (1.0-m) refractor at Yerkes Observatory. He took photographs of the Milky Way and discovered, together with Max Wolf, that the dark regions of the galaxy were clouds of gas and dust that blocked the light of the more distant stars. Barnard's niece Mary R. Calvert worked as his assistant and "computer" (at that time, this was the term used for people, usually women, who made calculations related to the astronomical work) from 1905 onward. He cataloged these dark nebulae and gave them numerical designations similar to the Messier catalog, from Barnard 1 to Barnard 370 (Fig. 2.8 shows Barnard [33] better known as the Horsehead Nebula). The initial form of this list was published in 1919 with a paper entitled "On the Dark Markings of the Sky with a Catalogue of 182 such Objects" in the *Astrophysical Journal*.

Barnard discovered in 1916 that the red dwarf star bearing his name (Barnard's Star) had a very large proper motion measured with respect to other stars. Barnard's star is the second closest star system to Earth, after the alpha (α) Centauri system.

E. E. Barnard died on February 6, 1923, almost 2 years after his wife Rhoda, who died in May 1921. Barnard passed away in Williams Bay, Wisconsin, and was buried at

Fig. 2.8. An example of one of Barnard's dark nebula, B33, better known as the Horsehead Nebula (Image courtesy of Don Taylor, www. skyshooter.net).

the Mount Olivet Cemetery in Nashville, Tennessee. His collection of photographs was published in 1927 in a work called *A Photographic Atlas of Selected Regions of the Milky Way*. This work was finished by Edwin B. Frost, the director of the Yerkes Observatory at that time, and Barnard's assistant Mary Calvert.

William Lassell

Our next person for consideration is William Lassell [38]. Lassell was born on June 18, 1799, in Bolton, England. By 1821, Lassell was observing with a 7.5-in. (19.1-cm) Gregorian reflecting telescope that he built himself. He was a frugal man with a good business sense. Even when he was financially well-to-do he still was conscientious of how he spent every penny, letting none go to waste. He began his successful business as a brewer in 1824, 2 years after finishing an apprenticeship in this line of work. When business was booming Lassell used the proceeds to build an observatory near Starfield, in Liverpool, which housed his 24-in. (61-cm) telescope, which was completed in 1845. It was in this observatory that he pioneered the use of an equatorial mount for easy tracking of objects as Earth rotates. In 1854, the entire observatory was moved out of Liverpool to Bradstones for better, clearer skies away from dirty industrial pollution (Fig. 2.9).

His frugality was displayed in his astronomical activities: with the 24-in. (61-cm) telescope he recorded his observations on second hand, "recycled" paper (that is,

Fig. 2.9. William Lassell and his observatory (Image courtesy of Mike Oates).

paper that has already been written on one side, but the other side is still blank). Yet he did show style, as exhibited in his earliest portrait made in 1845 that shows him in flashy attire; his Royal Astronomical Society presidential portrait shows him in a more fashionable (at that time) short jacket with braided collar (rather than the traditional, formal frock coat). It is interesting, and even a bit amusing, to note that Lassell recorded, in his (often homemade) notebooks, the exact position, in terms of angles of time, of the homes of many of his friends.

He spent time under the stars with his friends Alfred and Joseph King. They were entrepreneurs from Liverpool, England, by day and avid amateur astronomers by night, staying up most of the night to observe planets and stars and compare instrumentation (much as modern-day amateurs sometimes do). Lassell's friendship with these brothers almost certainly introduced him to his future wife Maria King, whom he married in 1827.

Lassell made a number of lifelong friends through his astronomical activities. These friends included William R. Dawes of Dawes' limit fame, the formula that describes the limit of resolution that a telescope or microscope is capable of. This formula was also credited to Lord Rayleigh. Dawes was a physician, astronomer, and non-conformist minister. Another friend was the engineer James Nasmyth. Nasmyth built a steam-driven grinding machine that was designed by Lassell to make 24-in. mirrors. Nasmyth likely made the heavy iron parts for the telescope.

Some notable observations that Lassell made included the discovery of Triton, the large moon of Neptune in 1846. This was discovered only 17 days after the planet itself was found by German astronomer Johann Gottfried Galle. Two years later, in 1848, he co-discovered Hyperion, which is the eighth known moon of Saturn. In 1851, he discovered two of Uranus's moons, Ariel and Umbriel. Around this time he sought clearer skies in Malta, where he observed regularly from that time on, as pollution from industrial activity in Liverpool, where he was living at the time, hampered observations. Lassell built his 48-in. (1.23-m) telescope in 1858, which was also used under the clear skies of Malta.

He knew the leading astronomers of his day, such as Otto Struve (of binary star fame and a very prolific astronomer with more than 900 journal articles and books to his name), who stayed with Lassell in 1850 at Starfield. Sir George Biddell Airy (of Airy disk fame) was also a guest of Lassell. Lassell and his work were highly regarded by Queen Victoria, who visited Liverpool in 1851, asking for Lassell by name and rose to her feet when he entered the meeting room. Also in that year, Lassell traveled to Sweden with another friend of his (a Mr. Stannistreet) to observe a total solar eclipse. During the event, the heat from the Sun concentrated by the powerful refracting telescope he was using onto a small spot on the dark glass (playing the role of Sun filter) in the eyepiece shattered the glass, nearly resulting in blinding Lassell. This is why, to this day, the use of the threaded dark glass solar filters is highly discouraged universally by astronomers.

In the mid-1860s, Lassell moved from Liverpool to Maidenhead in search of clearer skies in Britain. He remained there until his death in 1880, leaving the equivalent of several million dollars behind [39].

Edwin Powell Hubble

Our biographies would not be complete without describing the man who was instrumental in showing the world, for the first time, that the entire, unimaginably vast universe is expanding, and who is considered to be the father of modern cosmology [40, 41]. In fact, this man is so important that he has a space telescope named after him; the proportionality constant that relates the age and expansion rate of the universe is named for him; he has created a tuning fork diagram used to this day as a galaxy classification scheme; and there is also a website dedicated to his biography (see Appendix A in this book).

Edwin P. Hubble (Fig. 2.10) was born in the United States on November 20, 1889, in Marshfield, Missouri. His father was an insurance executive, John Powell Hubble, and his mother's name was Virginia Lee James. His family moved to Wheaton, Illinois, in 1900.

Fig. 2.10. Edwin P. Hubble (This item is reproduced by permission of the Huntington Library, San Marino, California).

Edwin was known more for his athletic abilities than his intellectual abilities in his early years, winning first place at a number of sporting events. For example, he won first place seven times and third place one time in a single high school track and field event in 1906. He also set, in that same year, the state record for the high jump in Illinois. In academics, he earned good grades in every subject except spelling.

After graduating high school, he attended the University of Chicago, where he studied mathematics, philosophy, and astronomy, and earned a Bachelor of Science degree in 1910. Three years prior to his death in 1913, Hubble's father moved the family from Chicago to Shelbyville, Kentucky, and ultimately near Louisville. Edwin was in England during his father's death, studying law at Oxford University, and he established a small practice in Louisville shortly after his return. Hubble taught high school for 1 year (his subjects were Spanish, physics, and mathematics, as well as coaching the boys' basketball team), then returned to his own studies of astronomy, earning a doctorate in astronomy at the University of Chicago in 1917. His dissertation was titled "Photographic Investigations of Faint Nebulae."

Hubble served in the U.S. Army in World War I, where he rose quickly to the rank of major. A year after the ending of the war, in 1919, Hubble took a staff position with George Ellery Hale at the Mount Wilson Observatory of the Carnegie Institution near Pasadena, California. He began a systematic, photographic study of galaxies and identified several examples of a special kind of variable star, called the Cepheid variable, which is used as a standard candle. These objects were found in several so-called spiral nebulae (including M31, the Andromeda Galaxy, and M33 in Triangulum). Despite opposition from the establishment of the time, including the renowned astronomer Harlow Shapley of Harvard University, who believed the Milky Way was the entire universe, Hubble published his findings first in *The New York Times* on November 23, 1924. This was followed by a formal presentation of a paper at the January 1, 1925, meeting of the American Astronomical Society. These findings fundamentally changed our view of the size and makeup of the universe.

It was at the 100-in. (2.5-m) Hooker Telescope at Mount Wilson Observatory where Hubble collected the observations of galaxies whose distances he measured based on Cepheid variables. He arrived at a value for the rate of expansion of the universe, a figure now known as the Hubble constant. He combined his own measurements of galaxy distances with the measurements of redshifts of these same galaxies made by Vesto Slipher and Milton L. Humason to discover a rough proportionality of the distance with the redshift of these galaxies. He started with 46 galaxies and obtained a Hubble constant of 500 km/s/Mpc, which is much higher than today's currently accepted value of 67.0 ± 3.2 km/s/Mpc (http://www.sciencedaily.com/releases/2011/07/110726101719.htm). The reason for the extremely high early value is the uncertainty in the distances to the galaxies and (to a lesser extent) their peculiar or random velocities. It was from the above-mentioned data that Edwin formulated, in 1929, what is now known as Hubble's law, which described how redshifted spectral lines and recessional velocity is related linearly. Not only does this conclusively show that the universe is expanding, it also demonstrates that it had a definite beginning some 2 billion years ago by Hubble's reckoning (we now are confident that this number is more like 13.7 billion years).

Hubble was not the only one to discover the expansion of space. Two years earlier, in 1927, a Belgian Catholic priest and physicist, Georges Lemaître, published the results of his own work in a not-so-well known Belgian, Annales de la Societe

Scientifique de Bruxelles. He showed that data that had been collected by Hubble and others was enough to derive a velocity-distance relation between galaxies, and this provided support for an expanding universe modeled on Einstein's General Theory of Relativity. This same Belgian physicist proposed the Big Bang theory of origins that same year, with observational support in the form of the observed velocities of distant galaxies along with the cosmological principle (a working assumption that observers on Earth do not occupy a privileged location or position).

Hubble devised a classification scheme to make sense out of these newfound "island universes," which we now have as the "Hubble tuning fork" diagram of galaxy morphology. Originally this diagram, which placed galaxies on the two-pronged tuning fork based on distance, content, shape, and brightness, was thought to be an evolutionary sequence, but later astronomers have shown otherwise. In the 1930s Hubble became involved in the geometry of the universe through studies of the distribution of galaxies throughout the cosmos. The data that he and others collected seem to indicate a flat, homogeneous universe, but at large redshifts there was a deviation from this flatness. Astronomer Allan Sandage provided his own explanation of what would become known as the Flatness Problem: "Hubble believed that his count data gave a more reasonable result concerning spatial curvature if the redshift correction was made assuming no recession. To the very end of his writings he maintained this position, favouring (or at the very least keeping open) the model where no true expansion exists, and therefore that the redshift 'represents a hitherto unrecognized principle of nature.'"

As it turns out, these deviations arose from the methodology with Hubble's survey technique, which did not account for changes in galaxy luminosity caused by the evolution of the galaxies themselves. In 1917, Albert Einstein's newly developed theory of general relativity showed that the universe was either expanding or contracting, but he did not believe this result. As a result, he introduced a "fudge factor," or cosmological constant, to avoid this "problem" of an expanding/contracting universe. After learning of Hubble's results concerning the expansion of the universe, Einstein realized that injecting this "fudge factor" was the largest blunder of his life.

Besides astronomy, Hubble was an activist, striving against the spread of fascism. Besides his service in World War I, he also served during World War II, where he came to the realization that he could make a more significant contribution as a scientist rather than as a soldier. He was given the honor of first use of the Palomar 200-in. telescope, which was completed in 1947. He suffered a heart attack in July 1949 while vacationing in Colorado, then passed away 4 years later, on September 28, 1953.

Hubble discovered the asteroid 1373 Cincinnati on August 30, 1935. Later he had another asteroid (2069) named after him, as well as the crater named Hubble on the Moon. He also is the namesake for the Edwin P. Hubble Planetarium, located in the Edward R. Murrow High School in Brooklyn, N.Y.; and the Edwin Hubble Highway, the stretch of Interstate 44 that passes through his birthplace of Marshfield, Missouri. The Edwin P. Hubble Medal of Initiative is awarded annually by the city of Marshfield, Missouri. The Hubble Middle School in Wheaton, Illinois, was renamed for Edwin Hubble when Wheaton Central High School was converted to a middle school in the fall of 1992. Edwin Hubble made the Hall of Famous Missourians 2003 and in 2008, and his likeness made the "American Scientists" U. S. stamp series (stamps were $0.41 that year).

Awards that Edwin Hubble received included the following: the Bruce Medal in 1938, the Franklin Medal in 1939, the Gold Medal of the Royal Astronomical Society in 1940, and the Legion of Merit for outstanding contribution to ballistics research in 1946.

The telescope that was named in his honor, the Hubble Space Telescope (HST), in many ways carried on his work of studying the expanding universe and pinning down the constant of universal expansion, the Hubble constant. It did so by locating and measuring the brightness changes of Cepheid variable stars in galaxies as far away as those in the Virgo Supercluster, some 65 million light years away. The HST was launched in 1990 and was soon found to have flawed optics, which were corrected in 1993 (more about the HST appears in Chap. 6). Since then, the HST has sent back hundreds of thousands of beautiful images of the universe in detail never before seen up until this time. The HST is still operating as of this writing (early 2012).

One of Edwin Hubble's quotes is: "Equipped with his five senses, man explores the universe around him and calls the adventure 'Science'." This is how many astronomers feel about exploring the universe through their amateur and professional astronomy.

Conclusions

We have just presented biographies of eight people (three from the same family, which shows how influential, astronomically speaking, this family was) who were important in advancing our knowledge and census of the deep sky objects we now love and enjoy. These are only a handful of a much larger crowd of people passionate about astronomy, whose work shaped the knowledge of the universe that we enjoy today. Indeed, there were literally thousands of contributors to the science of astronomy, from Earth's Moon to the most distant galaxies, but there are those relatively few who stand out.

Like a typical clear nighttime sky, looking up you see only a handful of really bright stars, a larger number of fairly bright stars, and a lot of faint, "ordinary" stars. People are like that also, but just like stars, no two people are exactly alike, and each is important in contributing to the light of the night sky (or the light of the knowledge of astronomy). The eight individuals described in this chapter lived mainly in the eighteenth, nineteenth, and early twentieth century; there are many others in these time periods, as well as the rest of the twentieth (and into the twenty-first) century, who made significant contributions to astronomy as we know it today.

Since the life and times of these pioneering astronomers, telescopes have grown larger and larger. The Mount Wilson 100-in. (2.54-m) telescope was the largest in the world from 1917 to 1948, when it was surpassed by the Hale 200-in. (5.1-m) telescope at the Palomar Observatory. The Hale telescope remained the largest until 1976, when the BTA-6 telescope took over as world's largest, at 6 m diameter for the primary mirror. This instrument is located at the Special Astrophysical Observatory in the Zelenchuksky district on the north side of the Caucasus Mountains in southern Russia. It remained the world's largest until the Keck Observatory took the title of largest telescope in 1993. With its 10-m mirror it remained "on top" as far as size is concerned until the Gran Telescopio Canarias became operational in 2009 with its 10.4-m mirror.

The next time you are outside with your telescope, either in the backyard or at your dark sky observing location, and you are hunting down these faint fuzzies, think about all the work that has gone into revealing the existence of the objects that you are viewing. This is work that, in many cases, spans more than a single century. Also, consider what we know about the physical nature of each individual object, information that only a short time ago was either unknown or otherwise largely kept from the general public. Consider the many and varied catalogs available today and how much work has gone into producing each, from observing and verifying the existence of each object to studying and learning more about that object. We truly live in a wonderful moment in history where the entire universe is, for the first time in history, opening up and revealing many of its secrets to us.

Chapter 3

The Nature of Star Clusters and Nebulae

Introduction

The next few chapters provide a basic "Astronomy 101" introduction to the types of objects that will be looked at later in this book (and you will look at later through your telescope). We are focusing on faint objects outside the local Solar System; information about faint planets, comets, and asteroids can be found in books written about these classes of objects. Our focus is on objects that can be found throughout interstellar space, along with objects found beyond our very own galaxy to the edge of the known and visible universe. But first here are some basics on the main kinds of deep sky objects that we know of.

We begin our survey of deep sky objects with two related classes of objects, star clusters and nebulae. The relationship is between two specific types of objects: open (or galactic) star clusters and bright (or emission) nebulae, and this will soon be described in more detail. Star clusters come in three kinds: stellar associations (clusters of stars whose members are not gravitationally bound together), open clusters, and globular clusters (in these two latter cases, the stars are bound gravitationally in the cluster). These types of objects are found in most every galaxy in the universe, but for the backyard observer, most objects of these types that can be observed are found in the Milky Way Galaxy. The brightest clusters are usually the nearest clusters, with more remote members obscured by distance, dust, or both.

Nebulae (plural for "nebula," or "space clouds") come in a variety of shapes, sizes, and types. Most of them appear quite faint in the nighttime sky due to their diffuse nature and relatively low light output. Notable exceptions to this include objects such as the Great Orion Nebula (M42) and the Lagoon Nebula (M8). Nebulae are found in most galaxies, but most nebulae accessible to backyard telescopes reside in our own galaxy. With some planning and preparation, however, you can locate nebulae in other galaxies (one of the brightest of which can be found in M33, the Pinwheel Galaxy in Triangulum, and has its own NGC designation: NGC 604).

B. Cudnik, *Faint Objects and How to Observe Them*, Astronomers' Observing Guides, DOI 10.1007/978-1-4419-6757-2_3, © Springer Science+Business Media New York 2013

Star Clusters

Some of the most spectacular objects in the nighttime sky are star clusters, both open and globular. True star clusters consist of a grouping of young or old stars that are gravitationally held together and can range in number from a handful (about 6 or 7) up to over one million.

Stellar Associations

These are not true star clusters in that the stars in the association are not bound by gravity. Eventually, within about ten million years or so, the member stars will escape each other's gravitational embrace and move on to different parts of the galaxy. Groups of stars that form a star cluster become an association when they no longer are gravitationally bound, or can start off as an association and stay that way if they fail to form an open cluster. Most of the stars of the famous Big Dipper are members of a former cluster, the Ursa Major moving group, and have similar proper motions. Other stars that seem unrelated are actually members of this moving group. Examples include Alphecca (alpha Coronae Borealis) and Zeta Trianguli Australis, both of which are related to the Ursa Major moving group.

Stellar associations are found within the arms of the Milky Way and other spiral galaxies. Associations in our own galaxy are easy naked-eye objects, with examples including the Perseus association (a group that surrounds Mirfak or α Persei), the Scorpius-Centaurus association, Cygnus OB2, and the Orion OB1 associations (some of these are mapped out in Fig. 3.1). These are also some of the easier objects to view in other nearby galaxies, as their OB associations appear as bright knots in the spiral arms of their hosts. Distant associations or moving groups cannot be readily identified because the distance itself makes their proper motions very difficult, at best, if not impossible, to detect.

Associations (at least the handful of nearest ones) are generally brighter groupings of stars that are best viewed with the naked eye or binoculars. They may also be seen in nearby galaxies such as the Magellanic Clouds and the Andromeda Galaxy, but it takes a telescope and a finder chart to locate these. Associations range in size from 100 pc, which is typical, down to a lower limit of 30 pc. This compares to a lower limit of true star clusters as 1 pc. Associations may occupy a more spherical volume of space, or they may be elongated; their shapes may provide an indication of their age. If an association is assumed to be spherical when its member stars form, then over time, the shearing forces of the Milky Way's rotation will stretch out the group, making it longer and longer until the individual stars disperse from the grouping and get "lost" among the member stars of the galaxy.

Associations are categorized into three groups. The first group is the easiest to observe: OB associations. These kinds of associations, as the name implies, are made up of primarily of young, bright, hot O- and B-type stars (to spectral type B2). The second group is the R-type association, which includes medium-mass stars. The Sun is thought to have been a member of the third type of association, the T-association, made up chiefly of T Tauri stars. T Tauri is a young pre-main sequent proto-stellar object that varies in brightness and serves as the prototype of these types of stars. These objects have similar masses and surface temperatures as late main sequence stars (spectral type F, G, K, and M) but are more luminous because of their larger sizes. They are objects in the process of contracting and

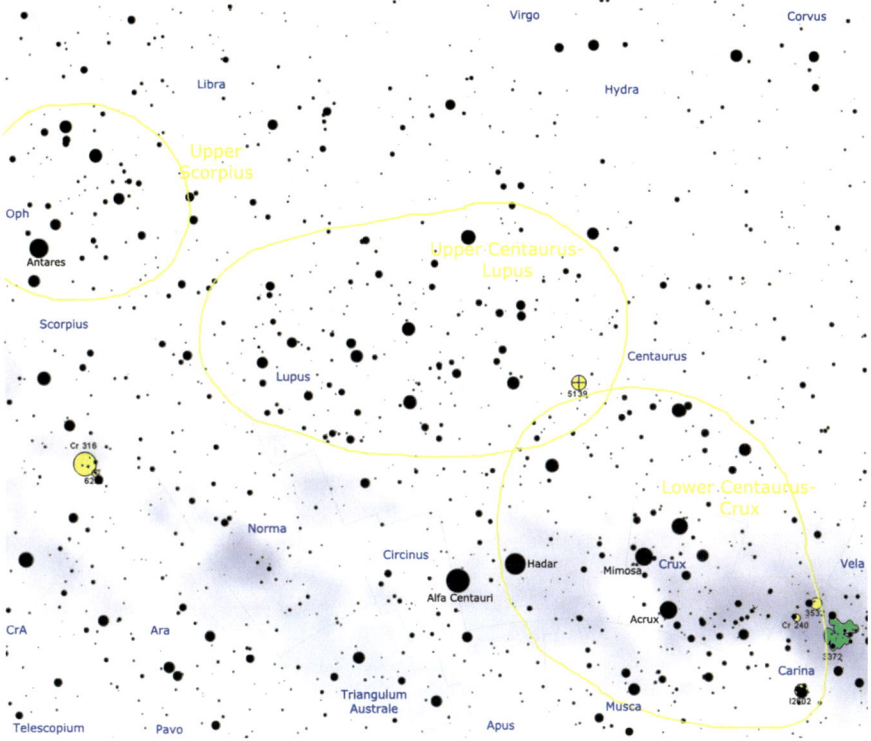

Fig. 3.1. Map of the Scorpius-Centaurus Association Complex (Image courtesy of Roberto Mura, CC-BY-SA 3.0).

have not yet begun the process of hydrogen fusion in their cores. Over time, the association of low-mass stars containing the Sun broke up and dispersed, with the Sun and its siblings traveling on their separate ways.

There are seven OB associations within 1,000 parsecs (pc or 3,260 light years/LY) of the Sun, and these include Perseus OB1, Orion OB1 (the nearest of them all, about 170 pc/550 LY distant), Scorpius-Centaurus, Monoceros OB1, Lacerta OB1, Cepheus OB3, and Cepheus OB2.

Open Star Clusters

There are many (more than 1,100 known, with the actual total possibly ten times more) open star clusters of all shapes and sizes that reside in the local galaxy and are observable through ground-based telescopes [42]. The best known are the Messier objects, such as M45 (the Pleiades), M44 (the Beehive or Praesepe), and M6 (the Butterfly cluster). There are also many obscure clusters that are either faint due to distance and/or extinction or blend in to the background Milky Way, making them difficult to observe.

Like stellar associations, open star clusters are groupings of stars that were formed together out of a common nebula. Unlike stellar associations, members of open star clusters are bound by gravity and generally stay together for a long time. However, over time, open clusters get disrupted by the gravitational influence of giant molecular clouds as both move throughout the galaxy. Although the individual members may continue to travel together for a time after the interaction, they may no longer be gravitationally bound and become a stellar association or a moving group. Although there are examples of young, middle-aged, and old open star clusters in our part of the galaxy, the vast majority of open clusters are young.

Both open clusters and stellar associations are thought to form out of a collapsing of part of a giant molecular cloud (GMC). This is a large, cold, dense cloud of gas and dust that can hold as much as many thousands of times the mass of the Sun and have densities that vary from 10^2 to 10^6 molecules of neutral hydrogen per cubic centimeter; star formation usually happens in regions of the cloud that contain densities above 10^4 H_2 molecules per cubic centimeter, which represents only about 1–10% of the cloud volume. Before they collapse, GMC's are "held up" in a sort of equilibrium state through magnetic fields, turbulence, and rotation. This equilibrium gets disrupted in many ways, including supernova shock waves, collisions with other clouds, and/or gravitational interactions. Once the collapse gets going, it begins to undergo fragmentation into smaller and smaller clumps that eventually form up to several thousand stars.

As these stars are forming, they are often hidden from view because of the dense, opaque material out of which they are made. Fortunately, modern astronomy has solved that dilemma. The material that makes up the cloud becomes transparent in the infrared part of the spectrum, and the young stars shine a lot more brightly in the infrared than the visible. Figure 3.2 shows two Hubble Space Telescope views of the Orion Nebula. The left view shows the optical spectrum image taken with Hubble's WFPC2 (Wide Field Planetary Camera), which reveals a few stars shrouded in glowing gas and dust. The right hand image was taken with Hubble's NICMOS (Near Infrared Camera and Multi-Object Spectrometer) infrared camera, which penetrates the glowing haze to reveal a swarm of stars as well as brown dwarfs.

Fig. 3.2. The central region of the Orion Nebula in visible light (*left*) and in infrared light (*right*) (Image credit: NASA and the Space Telescope Science Institute and the European Space Agency [ESA], Hubble ESA Information Centre under Contract NAS5-26555 [43]).

Just as there are physical processes that contribute to the formation of stars in a cloud, there are also processes that work against star formation. Only 30–40% of the gas in the cloud core actually forms stars. The hottest and most massive stars are the first to shine, and they put out copious amounts of intense ultraviolet (UV) radiation. This radiation ionizes the surrounding gas of the GMC, forming the visible "bright nebula" or HII region. Strong winds and significant radiation pressure will push away the hot, ionized gas at Mach 1; that is, at the speed of sound within the gas. The first supernovae in the cluster takes place in only a few million years as the most massive stars evolve fastest and are the first to undergo a core-collapse supernova event. The shockwave from this event will expel most of the gas in the region, effectively bringing a halt to star formation in that particular region. About half of the protostellar objects that are still present will remain, surrounded by circumstellar disks. Of these survivors, many of them will form accretion disks that could result in a planetary system.

One can find out whether an arbitrary grouping of stars is a star cluster by closely studying its members. Special Hertzsprung-Russell (H-R) diagrams called cluster color-magnitude diagrams help pinpoint cluster members, narrow down the age of the cluster, and determine its distance. A color-magnitude diagram plots the stars' colors (related to spectral type and temperature) versus their apparent magnitude to construct a main sequence and locate the giant stars in the cluster. Since H-R diagrams show specific patterns with the positioning of the main sequence stars and giants/dwarfs, stars that deviate from this pattern are likely not cluster members.

The offset of the observed main sequence, in terms of apparent magnitude, from a reference H-R diagram that shows a "model" case of stars plotted absolute magnitude versus color, gives the cluster's distance via the distance modulus formula. This formula is a mathematical expression, $m - M = 5\log_{10}(d/10)$, which uses the apparent magnitude (m or how bright it looks in the sky) of the star along with its absolute magnitude (M or how bright it would look if moved to a distance of 10 pc or 32.6 LY), to find out how far away the star is (the "d"). The point at which the main sequence turns upward and back to the right, the so-called "main sequence turnoff," gives an indication of the cluster's age. An example of a cluster color-magnitude diagram is shown in Fig. 3.3 and includes the stars from two different clusters (color-coded per cluster) on the same single plot.

It is assumed that all cluster members form at the same, or very close to the same time, in the history of the universe. It is also assumed that all the cluster members are at the same distance from Earth (which is permissible, since the physical size of the cluster is small compared to the distance). Some of the members are endowed with a lot of mass, some have less mass, and most have low mass. Stellar mass is measured in solar units, such that a star with two solar masses has twice the mass of the Sun, and so forth.

All stars shine by nuclear fusion, conveniently termed the "burning," of hydrogen at their cores to produce helium and lots of energy. The more mass, the greater the pressure and temperature at the core; the greater the pressure and temperature, the more vigorous the burning is; and the more vigorous the burning, the faster the star burns through its allotment of hydrogen and the shorter the lifespan of the star. In other words, massive stars live for relatively short periods of time whereas low mass stars live for a long, long time. The main-sequence turnoff on the cluster diagram shows which stars have burned through their allotment of

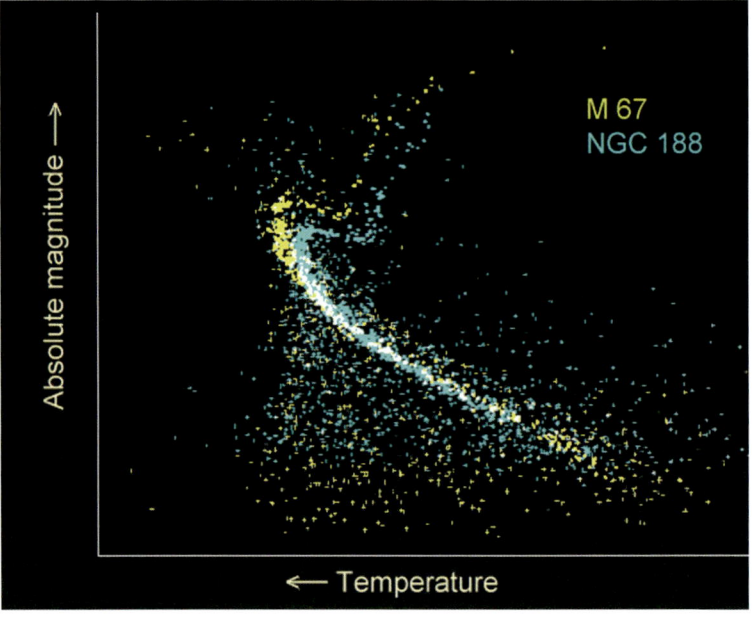

Fig. 3.3. Cluster magnitude diagrams for two open clusters. M67 stars are plotted with *yellow* points, and NGC 188 stars are plotted with *turquoise* points [44].

hydrogen and have "graduated" to "giant-hood"; this in turn gives a measure of the age of the star cluster.

Open star clusters (like the ones shown in Fig. 3.4) generally contain up to a few hundred member stars within a region up to 30 light years across. They are connected with nebula in that the cluster forms from the material making it up. Nebulae (more information about these is found in the section after "Globular Clusters") consist of huge clouds of gas and dust, the raw materials used for making stars. Both open clusters and nebulae are confined to the galactic plane and are almost always found within the spiral arms of the galaxy. There are many examples in the galaxy of clusters and nebulae in close proximity to one another. Brighter examples include the Lagoon Nebula, the Rosette Nebula, and the Orion Nebula. Images of these examples are shown in Fig. 3.5.

Another term for these examples of clusters associated with nebulae is "embedded cluster," which are star clusters that are partially or fully encased in interstellar dust and/or gas. The most famous embedded cluster is the Trapezium cluster in M44 (the Great Orion Nebulae). In the rho (ρ) Ophiuchi cloud (L1688) the core region has an embedded cluster.

As is the case when we are studying anything in the universe, we are interested in classifying the objects of our study, and open clusters are one such type of object. Open star clusters, which can be viewed as the fundamental building blocks of galaxies, come in different forms. These forms range from very sparse clusters, with only a few members, to huge agglomerations that each house thousands of stars. There is usually a prominent core (typically 3–4 light years across) surrounded by a halo of cluster members (extending out to about 20 light years from

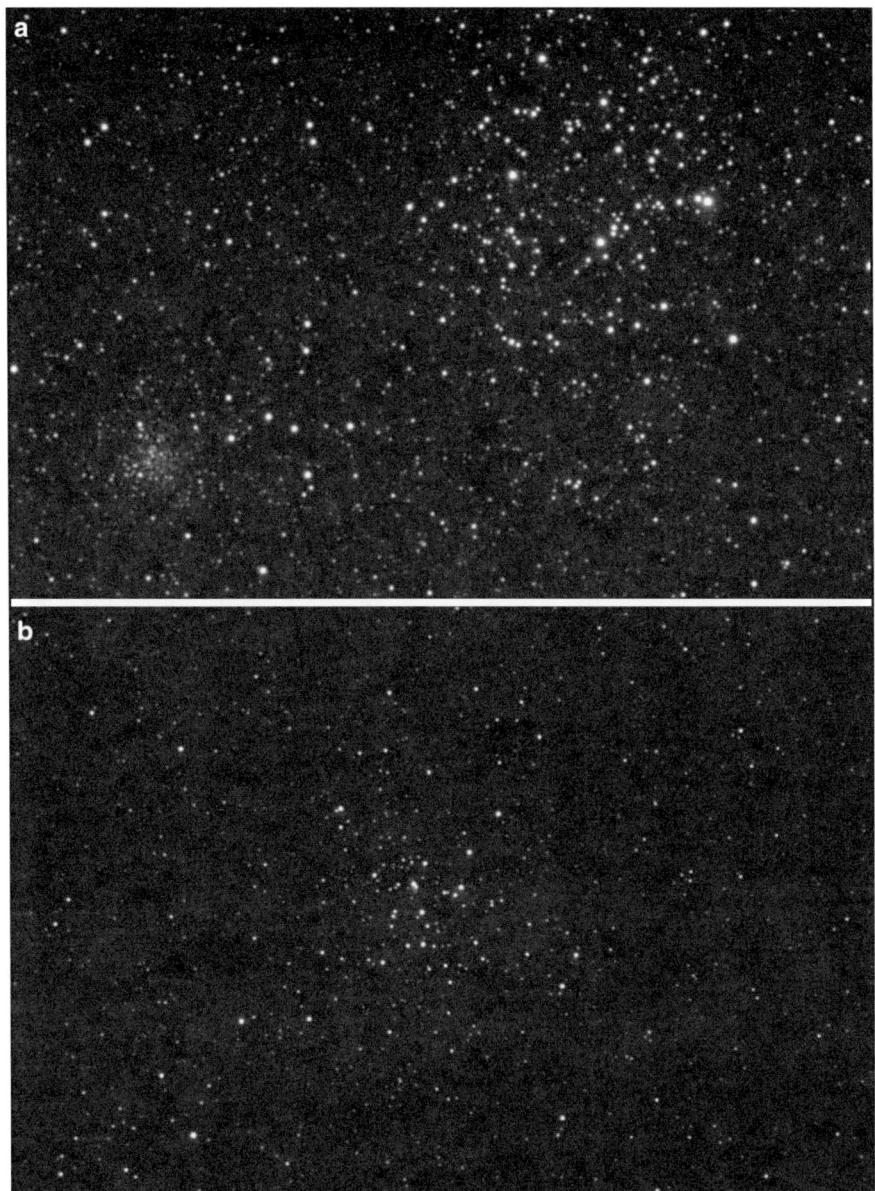

Fig. 3.4. Four examples of open clusters: (**a**) M35 (the larger of the two; the smaller is NGC 2158, which is four times more distant), (**b**) M36, (**c**) M37, and (**d**) M38 (Images courtesy of Loyd Overcash).

center). At the center of the cluster, the star density is about 1.5 stars per cubic light year; for comparison, the stellar density near the Sun is about 0.003 stars per cubic light year.

The astronomer Robert Trumpler developed a classification scheme for open clusters in 1930. He gave a three-part designation to a cluster, starting with a Roman numeral from I to IV to indicate the cluster's concentration and detachment

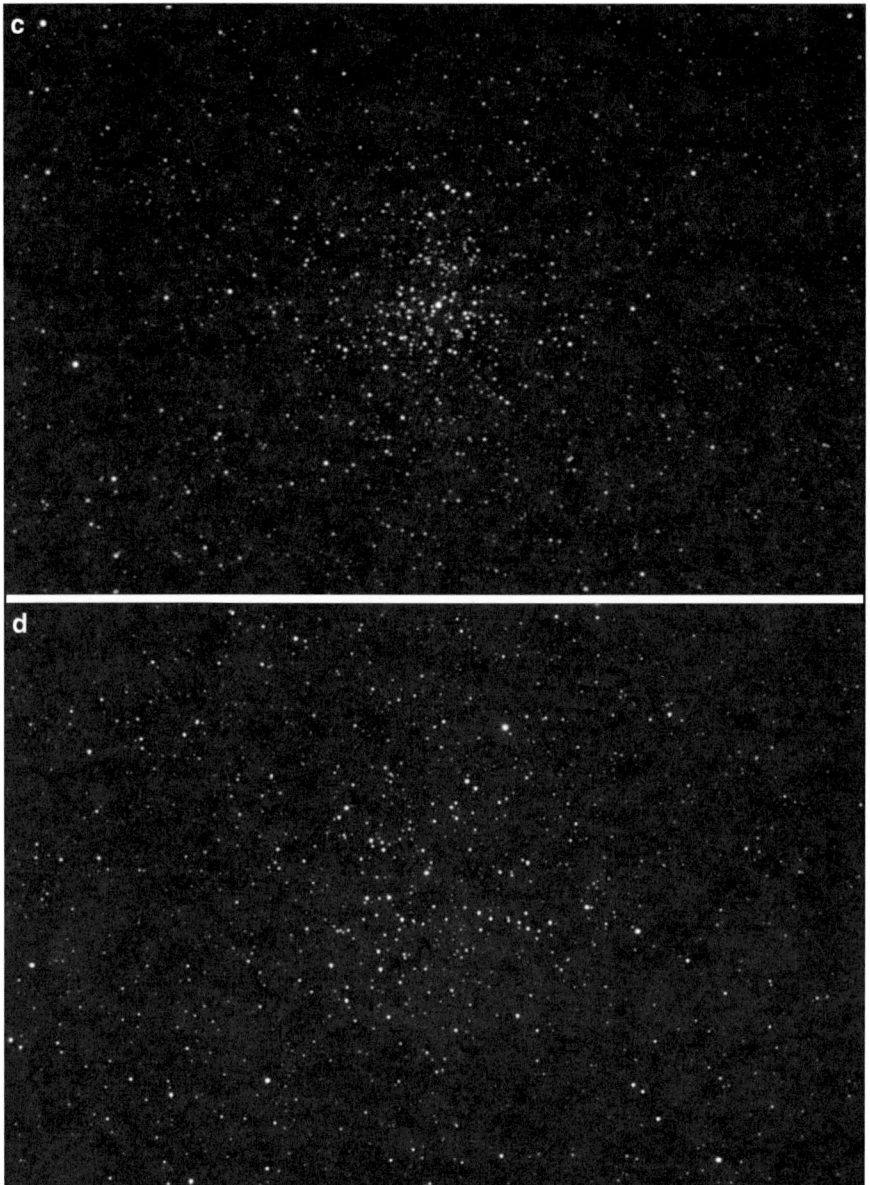

Fig. 3.4. (continued)

from the surrounding star field (from strongly to weakly concentrated), an Arabic number from 1 to 3 to indicate the range of brightness of members (from all members being the same or nearly the same brightness to having a wide range of brightnesses, some bright, some dim), across the same members of the cluster. Finally, the letters *p*, *m*, or *r* are given to show whether the cluster is poor, medium, or rich. If the cluster lies within nebulosity, the letter 'n' is appended.

Fig. 3.5. Four examples of cluster-nebulae associations: (**a**) NGC 6823, (**b**) M42, (**c**) M8 the Lagoon Nebula, and (**d**) the Rosette Nebula and its associated cluster (Images courtesy of Loyd Overcash).

Fig. 3.5. (continued)

Globular Clusters

Some of the most spectacular deep sky objects viewed through larger 'scopes are the globular clusters, which are roughly spherical collections of stars ranging in numbers from 10,000 to several million [45, 46]. Whereas open clusters tend to mostly occur near the plane of the Milky Way Galaxy, globular clusters are found mostly in the spherical halo of the galaxy. Some famous members include M13,

M22, M5, and M15 (all Messier objects). Although the Milky Way has 160 known globular clusters, Andromeda may have as many as 500 and the giant elliptical galaxy M87 may have as many as 13,000 globulars. Globular clusters orbit the galaxy at distances out to about 130,000 light years (40,000 pc) or more.

Globular clusters contain mainly very old Population II stars (which have a low proportion of elements other than hydrogen and helium, compared to the younger Population I stars like the Sun) that have been around for most of the history of the galaxy; these objects are very tightly bound gravitationally in a spherical volume of space several tens to hundreds of light years across.

The kinds of stars that are found in globulars are similar to those found in the bulges of spiral galaxies. Globular clusters tend to bunch up closer to the galactic center, which they orbit as satellites and which enabled the astronomer Harlow Shapley to determine the location of and distance to the center of the Milky Way in the 1910s and 1920s. He initially overestimated the distances to the clusters using their RR Lyrae variable stars, which he thought were Cepheid variables. RR Lyrae stars are intrinsically fainter than Cepheids. His distance estimate of the galactic center was also in error because of interstellar extinction (the dimming of stars due to dust between us and the stars), which was not taken into account at the time. However, Shapley's estimates of the size of the galaxy and the distance from the Sun to the galactic center were within an order of magnitude (less than a factor of 10) of the current accepted values.

The mechanism by which globular clusters are formed remains a poorly understood process. It is also not well understood whether the stars of a globular all form at once or in stages. However, age estimates of stars in many globulars indicate that most of the stars that make up each of these many are at approximately the same stage of evolution, suggesting they all formed at the same, or very close to the same time. But some clusters do show distinct populations of stars, one example being globular clusters in the Large Magellanic Cloud, which show that their stars were formed at two points in time. This may be due to an encounter with a GMC that triggered a period of star formation. The period of star formation is rather brief, compared to the age of the cluster.

The largest globular clusters, like Omega Centauri (Fig. 3.6a, along with other examples of globular clusters) and G1 (which belongs to the Andromeda Galaxy), begin to resemble dwarf spheroid galaxies, and some astronomers believe that some of these giant globulars may be cores of spheroidal galaxies that had been partially cannibalized by the Milky Way. These objects contain several million solar masses worth of stars and multiple stellar populations. It is estimated that about one-quarter of the Milky Way's globular clusters were "inherited" from dwarf galaxies absorbed by the Milky Way. There are several globulars, such as M15, that have very dense cores, and they appear highly condensed visually, indicating the presence of a black hole. Based on Hubble Space Telescope observations, M15 may host a 4,000 solar mass black hole, and Mayal II, a globular in the Andromeda Galaxy, may have a 20,000 solar mass black hole.

Globular clusters are bright enough (and far enough from the obscuring plane of the galaxy) to be visible from great distances. They are also bright enough to be seen around the nearest galaxies through backyard telescopes. The density of stars contained in a globular can be as high as 100–1,000 stars per cubic parsec (1.00 linear parsec is equivalent to 3.26 linear light years) at the core of the cluster, compared to an average of 0.4 stars per cubic parsec in interstellar space (in our part of the galaxy).

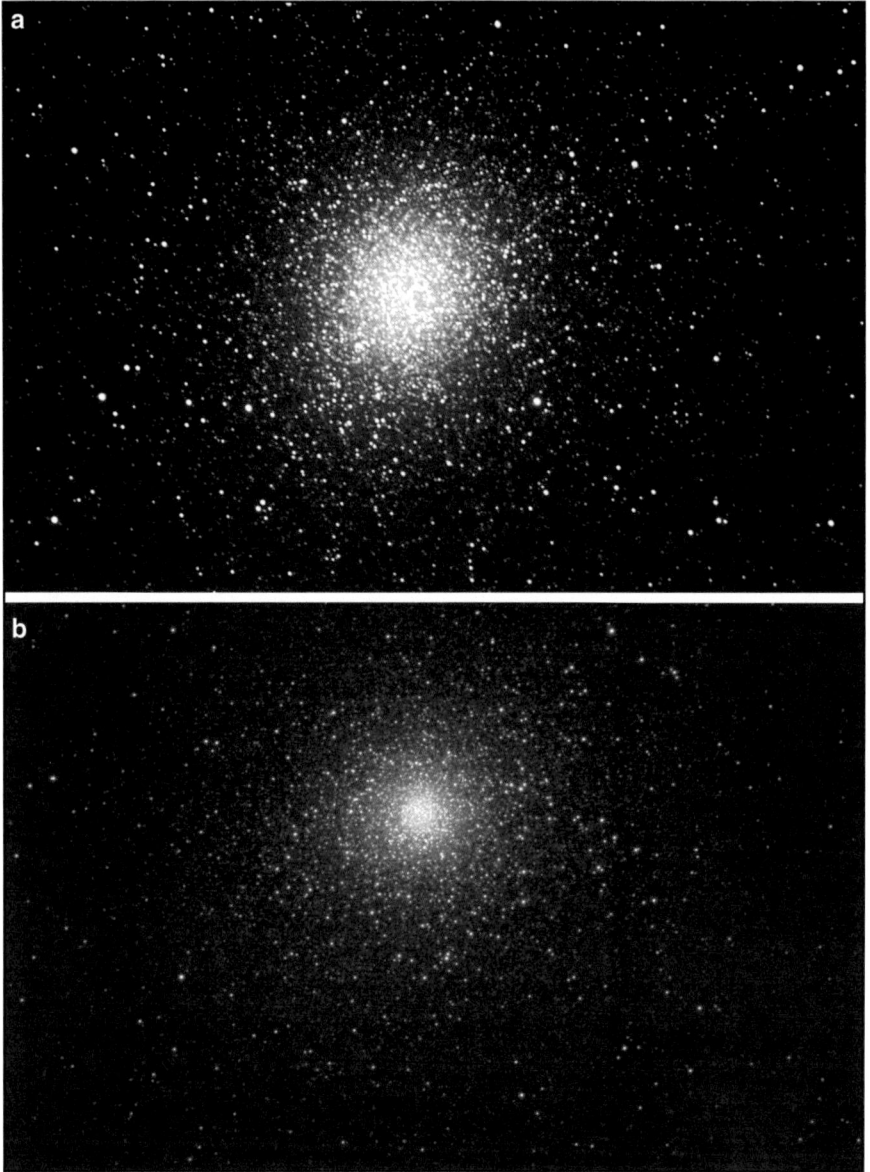

Fig. 3.6. Examples of brighter globular clusters: (a) omega Centauri, (**b**) M5, (**c**) M13, the Hercules Cluster, and (**d**) M3 (Images courtesy of Loyd Overcash).

Cluster color-magnitude diagrams have been plotted for many globular clusters as well as for open clusters as described above. Just like the case with the open clusters, these diagrams can tell us the distance and age of a cluster and can pick out cluster members from background (or foreground) stars. These diagrams also show that globular clusters in the Milky Way Galaxy tend to be old. Stars making up globular clusters are mostly yellow or red and less than two solar masses; more massive stars have long since exploded as supernovae or passed through their

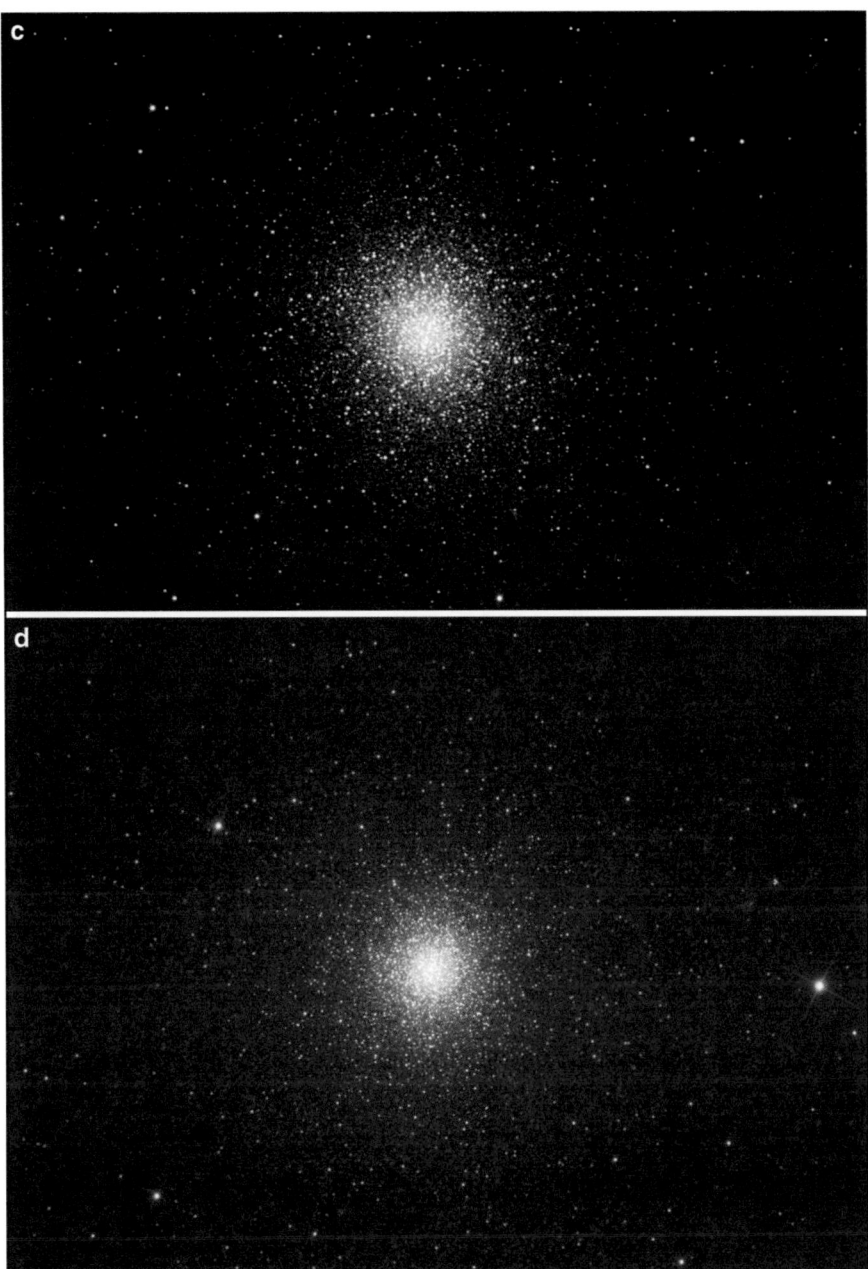

Fig. 3.6. (continued)

planetary nebula phases to become white dwarfs. There are rare blue stars called blue stragglers in globular clusters that are thought to have formed by stellar mergers in the dense inner regions.

The overall luminosities of the globular cluster population in the Milky Way and the Andromeda galaxies show that they can be modeled by Gaussian luminosity curves, similar to those used with planetary nebulae. These are also termed luminosity functions and can be used to figure out distances to other galaxies. The Gaussian or luminosity function can be represented with an average magnitude MV and a variance $\sigma 2$ (for the Milky Way Galaxy, $MV = -7.20 \pm 0.13$ and $\sigma = 1.1 \pm 0.1$). One can use this with the Milky Way or another remote galaxy by measuring the brightnesses of as many of its globulars as possible. Take these measurements and plot the number versus the luminosity on a graph, and fit a curve (Fig. 3.7 shows an example of such a galaxy, and Fig. 3.8 is what a luminosity function plot looks like; the example shown is for planetary nebulae, but the

Fig. 3.7. M104, the Sombrero Galaxy, as seen by the Hubble Space Telescope. Many of the star-like objects within the galaxy's halo are not stars but the brightest of this galaxy's many globular clusters (Image courtesy of NASA and the Hubble Heritage Team and STScI/AURA).

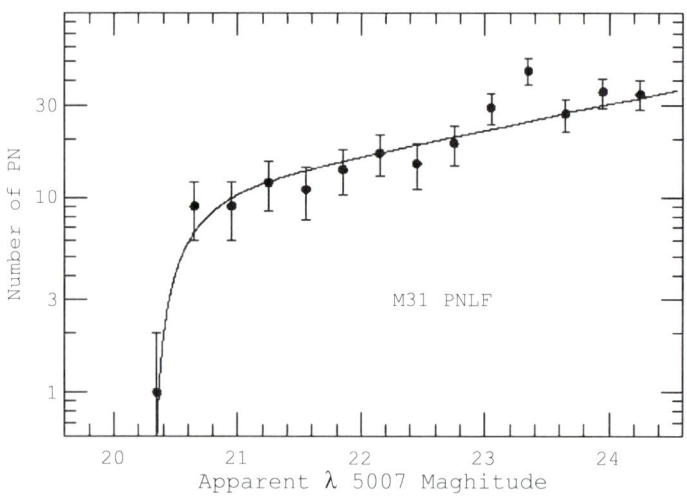

Fig. 3.8. An example of a planetary nebula luminosity function plot; this one is for M31 (Credit for the data and image: George Jacoby, NOAO/AURA/NSF).

process works similarly for globular clusters). The place where the curve drops can be used to determine how far away the globulars, and the host galaxy itself, are from us. The luminosity function is used as a standard candle to estimate distances to other galaxies.

Palomar 5 is a globular cluster that is being torn apart by the tidal forces of the Milky Way Galaxy. It is possible that the Milky Way had many more globular clusters near its beginning than the 160 that it has now. Most of these early globular clusters have long since been pulled apart, with their stars scattered throughout the galactic halo. Palomar 5 displays tidal tails; each tidal tail is more than 6,000 light years (1,800 pc) long or more than 10° in apparent length. These tails are quite useful for tracing out the stars' orbits. Palomar 5 currently lies about 60,000 light years (18,000 pc) above the plane of the Milky Way [47].

Two more globular clusters have been picked out of the obscuring dust and star clouds of the central regions of the Milky Way. The European Southern Observatory's VISTA telescope has imaged these previously unknown globulars (bringing the total from 158 to 160 known globulars in the Milky Way Galaxy) through obscuring material near the galaxy's center. These images are part of a systematic survey of the central regions of the Milky Way in infrared light. One of the two globulars, referred to as VVV CL002, may be the closest known globular to the galactic center [48, 49] (Fig. 3.9).

Until recently, there was a contradiction between the ages of globular clusters, based on theories of stellar evolution, which implied that they were older than the estimated age of the universe. The Hipparcos satellite made greatly improved distance measurements to globular clusters and these, combined with increasingly accurate measurements of the Hubble constant (with the Hubble Space Telescope), helped to refine estimates of the age of the universe (about 13.7 billion years), which is several hundred million years older than the globular clusters.

Morphology of globular clusters are done by means of standard radii. These include the core radius (r_c), the half-light radius (r_h), and the tidal radius (r_t). Overall, the luminosity of the cluster steadily decreases with distance from the center, and r_c is defined as the distance at which the apparent surface brightness has dropped by half. Since r_h contains stars in the outer part of the cluster that

Fig. 3.9. Two obscure GC's as viewed by VISTA (Image courtesy of the European Southern Observatory, D. Minniti, and the VVV Team [49]).

happen to lie along our line of sight with the core of the cluster, theorists will also define a half-mass radius (r_m), which is the radius from the core that contains half the mass of the cluster. Clusters that have an r_m that is small compared to the overall cluster size are thought to have a dense core. For example, M3 has an overall visible diameter of about 18 arc minutes, but an r_m of only 1.12 arc minutes.

The tidal radius of a globular is the distance from the center of the cluster where external gravitation from the galaxy has more influence over the cluster stars than the cluster itself. At this distance, individual stars can be removed from the cluster by the galaxy. For example, the globular M3 has a tidal radius, as seen projected on our sky, of about 38 arc minutes. This does not necessarily mean the stars stop at 38 arc minutes (there is a respectable number beyond the tidal radius), but outside that limit, the stars are subject to the gravitational tugs of the host galaxy.

Over time, globular clusters evolve by sorting out their stellar populations in a process known as "mass segregation." Because of gravitational interactions between individual stars, over time, single stars migrate from the center of the cluster to the outside of the cluster. This causes a net loss of kinetic energy from the core region that causes the stars left behind to occupy a more compact volume (they "bunch up" toward the center). This changes the overall surface brightness distribution, resulting in a brighter, more intense core.

Core collapse (of globular cluster stellar distribution) happens in three phases. During the cluster's adolescence, the core-collapse process starts with stars that are already near the core. Interactions between binary star systems hinder further collapse until the globular reaches its middle age. Finally, binary systems near the center get disrupted or ejected, leading to a tighter concentration of stars at the core.

The Hubble Space Telescope has provided observational evidence of this stellar mass-sorting process in globular clusters as more massive stars slow down and accumulate near the cluster's core, while less massive stars speed up and spend more time in the outer parts of the cluster. The globular cluster 47 Tucanae, second most massive Milky Way globular cluster known after ω Centauri, has about one million stars and is one of the densest globulars in the southern hemisphere. Astronomers were able to determine precise velocity measurements for almost 15,000 stars in this cluster.

Intermediate Forms and Star Super Clusters

Objects such as Westerlund 1 in the Milky Way may be the precursors of globular star clusters. Some of the Milky Way's 160 globular clusters may have been captured from small galaxies disrupted by our galaxy (M79 may be one such example). An object known as BH 176 in the southern part of the Milky Way shows properties of both an open and a globular cluster, which blurs the distinction between the two types of cluster. This seems to be a completely new kind of star cluster, an example of which was found in the Andromeda Galaxy in 2005. The object is like a globular cluster but much larger (several hundred light years across) and hundreds of times less dense. The distances between the stars are much greater as a result of the significantly lower density. These objects lie somewhere between a globular cluster and a dwarf spheroidal galaxy. Aside from Westerlund 1, there are no similar types of intermediate clusters known in the Milky Way. Examples in M31 are M31WFS C1, M31WFS C2, and M31WFS C3.

Some properties of these intermediate type or super clusters include contents of hundreds of thousands of stars, old (Population II) stellar populations, and metallicity. The idea of metallicity in stellar astronomy takes into account the concentrations of hydrogen and helium while lumping everything else into a category called "Metals," and it is often used as an indicator of the relative age of something. If an object like a star or nebula contains a relatively higher proportion of metals, then it is thought to be a relatively young object, since, over the course of the history of the universe, many generations of stars are thought to have produced the metals in their cores via nuclear fusion. The metals "get out" when the star becomes a planetary nebula or "goes supernova."

Nebulae

A nebula is an interstellar cloud of dust, hydrogen, helium, and other ionized gases that is often a region of star formation. Nebulae are classified according to their physical nature: emission, reflection, planetary, supernova remnant, and dark. These different types reflect stars in different stages of their lives, from their births through their deaths and in between. Dark nebulae, like Barnard 33, are mainly opaque, non-illuminated clouds of gas and/or dust seen in silhouette against more distant star clouds or nebulae. American astronomer E. E. Barnard (whose biographical sketch appeared in the last chapter) was instrumental in cataloging these dark nebulae.

Emission Nebulae

Emission nebulae are bright nebulae that glow because of their being irradiated with copious amounts of UV light from nearby hot stars [50]. These hot stars may be the most massive members of the stars formed from the gas/dust or nearby stars that are massive enough and near enough to irradiate and ionize the contents of the nebular cloud. These objects serve as nurseries where stars are being formed out of the nebula's raw materials. Examples of such objects include the Swan Nebula (M17, Fig. 3.10a), the North America Nebula (NGC 7000, Fig. 3.10b), the Orion Nebula (M42), and the Rosette Nebula (NGC 2237-9).

(Portions of this chapter and the next two, as indicated, are derived from SEEDS. *Stars and Galaxies*, 6E. © 2008 Brooks/Cole, a part of Cengage Learning, Inc. Reproduced by permission.) [51]

The so-called "forbidden" spectral lines were noticed in nebulae when spectroscopy began to be widely used in the nineteenth century. These lines are called "forbidden" because they are not seen under ordinary laboratory conditions on Earth's surface. The probability of these types of energy transitions that produce these specific lines are extremely low. However, in the extremely low density gaseous environment of nebulae conditions are favorable for these transitions to happen, producing these special spectral lines. A pair of examples of these special lines (and a special kind of filter, important for looking for faint versions of these objects, is based off of these) is the OIII (oxygen with two of its electrons taken off) pair at 495.9 and 500.7 nm (the green part of the rainbow of visible light colors).

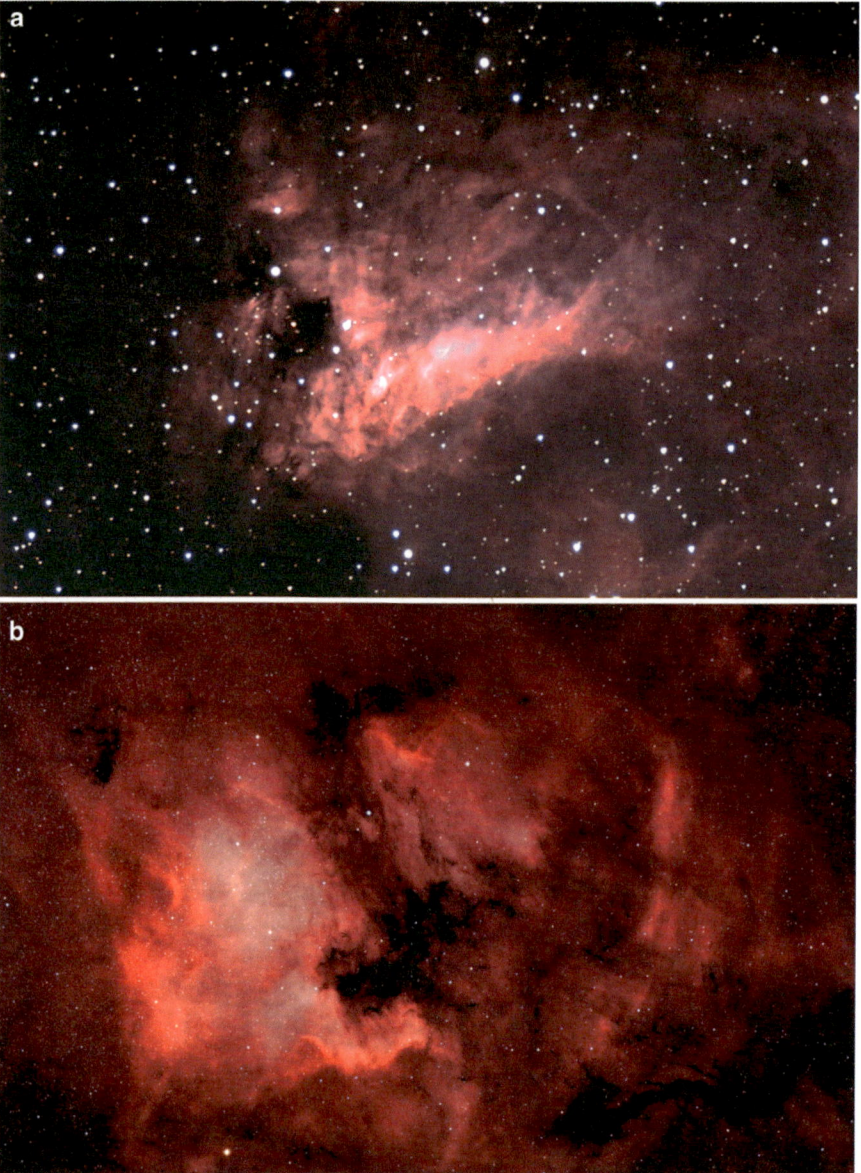

Fig. 3.10. Two examples of emission nebula: (**a**) the Swan (Image courtesy of Paul and Liz Downing), and (**b**) the North America & Pelican Nebulae (Image courtesy of Don Taylor).

Emission nebulae come in a wide variety of sizes, shapes, and colors and are observed in most spiral and irregular-type galaxies as well as in our own Milky Way Galaxy. Because hydrogen makes up most of the interstellar gas, and because hydrogen has relatively low ionization energy, many emission nebulae appear red (in images) due to the strong emissions of the Balmer series of lines. Nebulae would appear green and blue if more energy were available to ionize other elements

Fig. 3.11. Two examples of narrowband images of emission nebulae: (**a**) the North America and Pelican nebulae (compare with Fig. 3.10b), and (**b**) the Elephant Trunk Nebula region (Both images courtesy of Don Taylor).

that make up the objects. Some of the most beautiful of objects as imaged by the HST as well as ground-based observatories (both amateur and professional) are emission nebulae. Visually the object, save for the brightest ones, may be less than appealing, but to the trained eye each specimen appears different from the next.

Figure 3.11 provides examples of narrowband images of emission nebulae. These images are color-coded to trace out where large amounts of gaseous species such as hydrogen (in the form of H-alpha), oxygen (twice ionized oxygen, or OIII), and sulfur (once-ionized sulfur, or SII) are located within a nebular complex. These images are made not only with HST observations and professional observatories but with more and more amateur 'scopes, which are creating scientifically useful narrowband images of nebulae. For more information on these types and technologies of imaging, consult Appendix B of this book, which has a listing of resources for further study.

Reflection Nebulae

Reflection nebulae shine by reflected starlight from stars that are usually not luminous enough to produce the extreme amounts of UV necessary to make emission nebula glow. In addition, the reflection nebulae are made up of either gas or dust (or a mix of both).

One of the brightest of the 500 or so that are known to exist in the Milky Way Galaxy is M78 (Fig. 3.12a) in the constellation Orion. The famous case of McNeil's Nebula in 2004 was a reflection nebula that varied in brightness because of an outburst of a very young stellar object in the process of forming. This is one of a handful of nebulae that are known to change in brightness due to their parent or illuminating star(s) varying in brightness. Other examples of variable nebulae include Hind's Nebula (NGC 1554/5), Hubble's Variable Nebula (NGC 2261), NGC 6721 in Corona Australis, and Gyulbudaghian's Nebula. A photogenic mixture of emission and reflection nebulae show up in the region of rho Ophiuchus, as depicted in Fig. 3.12b. There are a total of about 500 reflection nebulae known.

The dust within the nebula scatters starlight and carries a reflection of the spectrum of the source star [52]. Reflection nebulae appear blue because photons of shorter wavelengths scatter more than photons of longer wavelength do, just like what happens in the daytime sky on Earth. The dust particles are quite small and so are able to scatter shorter wavelength photons better than longer wavelength photons. Often, reflection and emission nebulae are seen together and are sometimes both referred to as diffuse nebulae.

The Pleiades (M45) show another example of a reflection nebula originally thought to be the leftovers of the raw material that went into making the member stars of the Pleiades. It is now likely that the nebular wisps are actually material from the interstellar medium through which the cluster is passing, causing the feathery cirrus-like texture of the denser interstellar medium (the dusty component) material to show up.

Planetary Nebulae

Planetary nebulae are emission nebulae that are remnants of low mass and Sun-like stars that have ended their lives, shedding most of their material into space, leaving behind a glowing shell of ionized gases and a core, which becomes a white dwarf star [53]. These nebulae are so named due to their superficial resemblance to planets in telescopes, but rather than being solid spheres or gas giants, these objects are expanding, fluorescing, tenuous clouds of gas made to shine by the ionizing radiation from their central stars. Examples of planetary nebulae include M57 (the Ring Nebula in Lyra), M27 (the Dumbbell Nebula in Vulpecula), M76

Fig. 3.12. (**a**) M78 and its environs (Image courtesy of Loyd Overcash), and (**b**) the rho Ophiuchi region (Image courtesy of Don Taylor).

(the Little Dumbbell in Perseus), and NGC 7027 (the "Peanut" in Cygnus). Several of these are shown in Fig. 3.13.

Planetary nebulae are far more complex than what the visual views through the eyepiece would have us believe [54]. A careful study of the fine details of planetary nebulae as seen in images reveals clues about the physics of the dying stars' last "days." At the end of a star's life, the star's outer layers get expelled as a result of pulsations and stellar winds. As these layers expand outward, the hot luminous

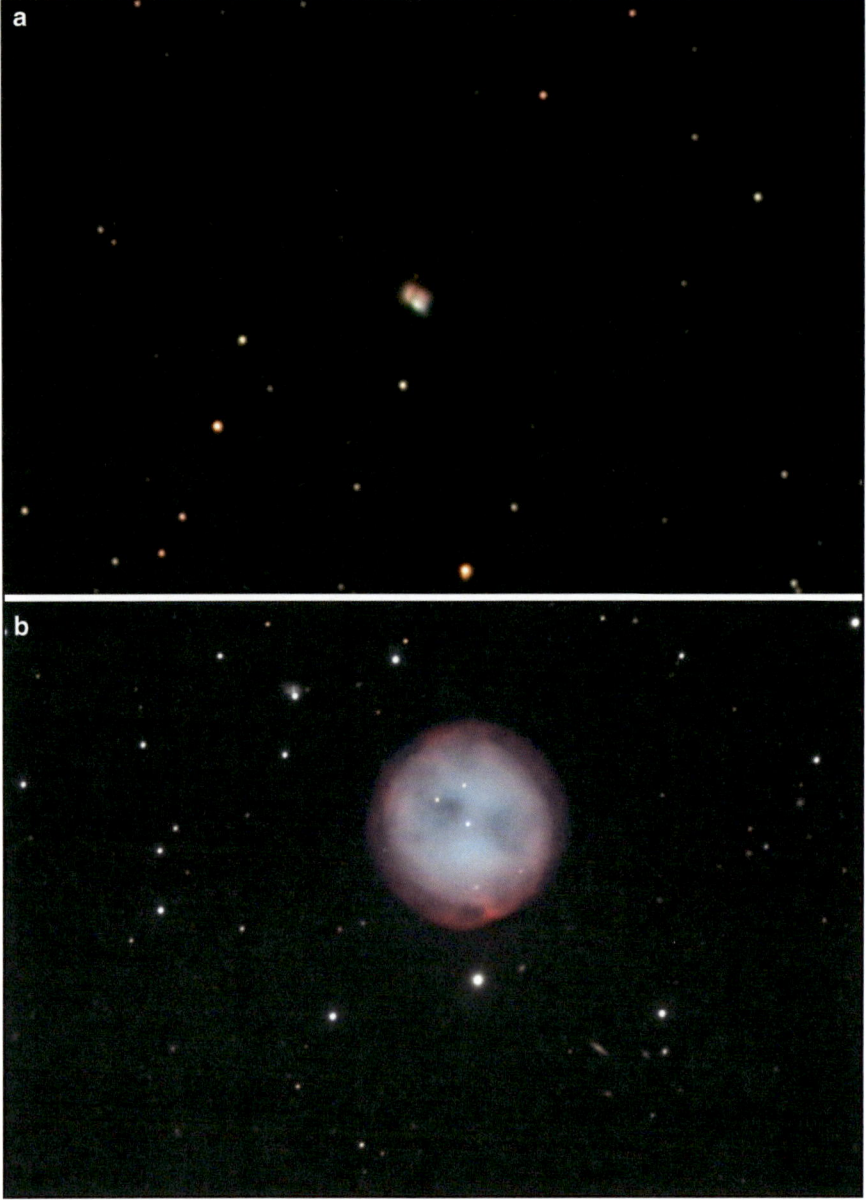

Fig. 3.13. (**a**) NGC 7027 the "peanut", image courtesy of Paul and Liz Downing; (**b**) M97 the Owl Nebula, image courtesy of Loyd Overcash; and (**c**) NGC 246 Planetary Nebula, image courtesy of Loyd Overcash.

core of the star gets exposed, and it undergoes a two-step evolution. It starts by getting hotter as it contracts, generating hydrogen fusion in a shell around the core. It maintains a constant luminosity, getting hotter and hotter while shrinking in size. The central star reaches temperatures around 100,000 K and gives off copious amounts of ultraviolet radiation that ionizes the expelled layers, making the planetary nebula visible.

Fig. 3.13. (continued)

Oppositely directed jets in the system produce many of the asymmetries that are seen in images of planetary nebulae. Eventually the central star cools down into the white dwarf phase, the second step of this evolution, and as it does so, the amount of UV radiation decreases. Ultimately the levels of UV are no longer high enough to ionize the increasingly distant gas cloud, and it becomes invisible from Earth, effectively ending its planetary nebula phase of evolution.

HST images of planetary nebulae, like emission nebulae, are some of the most beautifully striking images produced by the orbiting observatory. These show the intricate features, just mentioned, that reveal the dynamics of the system and even whether the system was a binary star system or not. About one-fifth of the imaged planetary nebulae are roughly spherical, but most are not spherically symmetric. The shapes and features arise from mechanisms not well understood as of yet, but binary central stars, stellar winds, and magnetic fields may all play a role in the shaping of the nebula. The gases of planetary nebulae contained a high proportion of elements such as carbon, nitrogen, and oxygen, and these are carried away into the interstellar medium as these gases expand and disperse. In so doing, these objects enrich the ISM with heavier elements essential for planet-building.

Planetary nebulae tend to have high surface brightness, but they only last a few tens of thousands of years. A typical nebula is roughly one light year across and is made up of extremely rarefied gas (density of 100–10,000 particles per cm^3). Because of their high surface brightness, through a large telescope, the visual appearance of some of these objects actually displays some of the finer details imaged professionally. Color shows up for some objects even through smaller instruments and can range from bluish to emerald green (depending on the object itself and the color sensitivity of the observer's eyes). Fainter objects all appear grayish in color visually. We know of about 3,000 planetary nebulae in our galaxy that are found mostly near the plane of the galaxy and with the highest concentration toward the galactic center.

One common perception is that all low mass stars form planetary nebulae when they die, but this is turning out to not quite be the case. In order for a visible planetary nebula to form, the core (which becomes the white dwarf) needs to have a surface temperature of at least 25,000 K to ionize the ejected gases to make them glow. The Sun may be on the borderline and may not even get hot enough as a white dwarf for this to happen. Certainly stars with masses less than the Sun will not get hot enough for planetary nebulae to form, and stars with less than 0.8 solar masses have not had enough time in the history of the universe to evolve through the red giant phase.

Planetary nebula haloes were ejected by increasingly unstable "Asymptotic Giant Branch" red giant stars prior to the formation of the main nebulae and were only discovered relatively recently. There are many planetary nebulae whose extended haloes may be observable with large backyard 'scopes from very dark skies. These outer haloes provide a special challenge to find faint components to bright field objects. Table 3.1 is the "Selected Planetary Nebula Outer Halos" list, courtesy of *Sky & Telescope* magazine [55]. The best way to see planetary nebula outer haloes is to use an OIII filter and observe under dark, clear, transparent skies.

Table 3.1. Selected PN outer Halos, courtesy sky and telescope magazine [55]

Object	Constellation	RA (2000.0) h m s	Dec.° '	V mag.	PN Diam.	Halo visual diameter[a]	Halo notes[b]
NGC 6210	HER	16 44 30	+23 48.0	8.8	14″	70″	Diffuse. Brighter on its north and west fringes
NGC 6543	DRA	17 58 33	+66 38.0	8.1	18″ × 30″	>54″	Knot in halo 1.5' west is IC 4677. Photographic halo is 4' wide
M57	LYR	18 53 35	+33 01.7	8.8	71″	140″	Very difficult; not seen with 10-in
NGC 6826	CYG	19 44 48	+50 31.5	8.8	25″	75″	Slight variations in surface brightness. More prominent on the eastern side
M27	VUL	19 59 36	+22 43.3	7.3	350″	700″	Needs large aperture. Knots in halo visible
NGC 6891	DEL	20 15 09	+12 42.3	10.5	14″	42″	Requires moderately high magnification. Check by blinking with OIII filter
NGC 7026	CYG	21 06 19	+47 51.1	10.9	20″	60″	Bipolar halo with lobes extending NW and SE. Visible without OIII filter
NGC 7662	AND	23 25 54	+42 32.1	8.3	12″	36″	NE area more prominent. Nebula's disk blends into halo
M76	PER	1 42 20	+51 34.5	10.1	65″	130″	Bright and noticeable without OIII filter. The "wings" of the Little Dumbbell
IC 418	LEP	5 27 28	−12 41.8	9.3	12″	84″	Without OIII filter halo extends 30″ or 40″
NGC 2371/72	GEM	7 25 35	+29 29.4	11.3	55″	165″	Broken arcs, difficult to discern, are SE and NW of poles
NGC 2440	PUP	7 41 55	−18 12.5	9.4	14″	75″	Bipolar halo requires large aperture and magnification to show structure
NGC 3242	HYA	10 24 46	−18 38.5	7.8	40″	>120″	Easy even without OIII filter. Filament 12' WSW needs large aperture

[a] Values are approximate, size may vary with aperture

[b] Observing notes were made with a 10-in. f/5.5 reflector and, unless otherwise specified, an OIII filter

The guidelines for faint objects (in terms of 'scope preparation, sky conditions, etc.) are especially applicable for planetary nebula outer haloes.

Planetary nebulae have been found as members of four globular clusters (M15, M22, NGC 6441 and Palomar 6). On a tour of Lowell's 24-in. (0.61-m) Alvin Clark telescope, the instrument was pointed at M15 and someone provided an OIII filter with which we "blinked" M15 to see its planetary nebula. This was the author's first experience at "blinking" planetaries, and it really works: the stars of the globular cluster are markedly dimmed when the filter is passed into the line of sight, while the stellar planetary nebula remains near its original brightness.

Although planetary nebulae have been seen among the stars of globular clusters, none has been conclusively found to be a member of an open cluster, even though two cases of planetary nebulae superimposed in front of open star clusters are known. These cases include NGC 2348 seen in front of the stars of M46, and NGC 2818 seen in front of an open cluster by the same designation. Open clusters have much less gravitational cohesion than globular clusters, and most open clusters disperse after 100–600 million years.

Although it is possible to have one or more planetary nebula be a member of an open cluster, there are a variety of factors that limit our chances of finding one. For example, the planetary nebula phase for more massive stars belonging to younger clusters lasts only a few thousand years. Another reason is that stars that do pass through a planetary nebula phase that are closer in mass to the Sun will have done so after its host cluster has dispersed.

Interstellar Dust

This material is spread throughout interstellar space and makes up about 1% by mass of the interstellar medium [52, 56]. Dust is responsible for the dark lanes that are seen in the summer Milky Way (as viewed from a dark-sky location) and for the dark splotches seen in images of other galaxies. Dust also shows up visually in telescopic views of M31 (the Andromeda Galaxy), M104 (the Sombrero Galaxy), M64 (the Blackeye Galaxy), and NGC 4565 (a beautiful edge-on spiral galaxy). Closer to home, in our own Milky Way, dust is responsible for an effect known by astronomers as interstellar extinction – the dimming of distant stars by a factor of one magnitude per one thousand parsecs. It also is responsible for interstellar reddening, which makes blue stars look redder. This comes from very tiny dust particles scattering starlight. These dust particles are so tiny they are about the same size as the wavelength of visible light photons. Because of this, blue light is scattered more readily than red light.

Interstellar space is filled with regions of lower and higher density gas in various places, some of which show up as either emission or reflection or dark nebulae. What we actually see in the eyepiece are small parts of the interstellar medium made visible, either directly by glowing (emission and reflection nebulae) or indirectly by silhouette (dark nebulae, "Bok globules") as the material blocks light from more distant sources. Astronomers can use longer wavelength light, such as infrared light, to probe the galaxy; and 21-cm radiation is used to map the galaxy's structure since it is readily available and able to pass through most of the interstellar medium. A wide variety of molecules have been found in the interstellar medium, which include the following: hydrogen sulfide, ethyl alcohol, cyanogens,

Table 3.2. Selected molecules detected in the interstellar medium (SEEDS' table 10-1 [56])

H_2	Molecular hydrogen
H_2S	Hydrogen sulfide
C_2	Diatomic carbon
N_2O	Nitrous oxide
CN	Cyanogens
H_2CO	Formaldehyde
CO	Carbon monoxide
C_2H_2	Acetylene
NO	Nitric oxide
NH_3	Ammonia
OH	Hydroxyl
HCO_2H	Formic acid
NaCl	Common table salt
CH_4	Methane
HCN	Hydrogen cyanide
CH_3OH	Methyl alcohol
H_2O	Water
CH_3CH_2OH	Ethyl alcohol

methyl alcohol, formaldehyde, hydrogen cyanide, acetylene, and methane, to name a few. Table 3.2 provides a sampling of the kinds of molecules that have been found in the interstellar medium.

Giant molecular clouds are the birthplaces of stars and form some of the material for some types of deep sky objects. One example of faint deep sky object are the Herbig-Haro objects, where a newborn star emits powerful jets of material along its polar axes; when these jets hit the ISM, they become visible. The Elephant Trunk Nebula, a globule of dark nebulae, consists of material compressed and twisted by radiation and winds from a nearby luminous star. Infrared images of this region show six protostars within the dark nebula. The Eagle Nebula is shaped by radiation and winds from massive stars. Erosions of these "pillars of creation" have exposed small globules of denser gas and dust, about 15% of which have formed protostars. EGG, *Evaporating Gaseous Globules*, is the name given to these.

Supernovae Remnants

Supernovae remnants (such as the examples in Fig. 3.14) are leftovers of massive stars that ended their lives in huge explosions called supernovae ("supernova" is the singular form) [57]. The remnants, bound by expanding shock waves and containing material ejected during the supernova explosion, are made visible as they expand into the interstellar medium, colliding with the ISM's contents, causing the gas to glow. These objects come in a variety of forms, from the patch-like Crab Nebula (M1 in Taurus) to the stringy Veil Nebula (part of the Cygnus Loop that includes NGC 6960, 6979, 6992, 6995). In many cases, studies of the expansions and motions of these objects pinpoint the length of time that had passed since the supernova explosion took place.

Fig. 3.14. Examples of supernova remnants: (**a**) M1, the Crab Nebula, (**b**) NGC 6960 part of the Veil Nebular Complex (Images "a" and "b" courtesy of Paul and Liz Downing), (**c**) NGC 6888, the Crescent Nebula (Image courtesy of Loyd Overcash).

There are two broad types of supernovae: a massive star at the end of its life that collapses inward under the force of its own gravity to form a neutron star or black hole; or a white dwarf in a binary system where the white dwarf is accreting material from a companion star until it reaches a critical mass and suffers a thermonuclear explosion that destroys the star. In both cases, material gets ejected at up to 10% the speed of light, or 30,000 km/s, and a strong shock wave forms ahead of the

Fig. 3.14. (continued)

ejecta. This shock wave heats the upstream plasma to temperatures up to millions of Kelvins. The shock wave slows down over time as it sweeps up material from the medium, but it is able to expand over hundreds of thousands of years, traveling tens of parsecs before it slows below the local speed of sound.

SN 1006 is 1,005 years old and 60 light years across [58]. It was produced by a type Ia supernova. The Cygnus Loop is between 5,000 and 10,000 years old and 80 light years (25 pc) across. The supernova remnant Cassiopeia A is around 300 years old and about 10 light years (3 pc) across. This object is the product of a Type II supernova and contains a neutron star. The Crab Nebula, M1, is the remains of a supernova that exploded in A.D.1054. The glow visible through the eyepiece is produced by a phenomenon called synchrotron radiation, where electrons release visible light as they spiral around magnetic field lines that permeate the nebula. The filaments of M1 can be seen with an OIII filter visually through a 14-in. Cassegrain; these are caused by ionization of the gases by the intense radiation from the Crab Nebula pulsar.

There are three types of supernova remnants: shell-like remnants, such as Cassiopeia A; composite remnants that include a shell with a central pulsar wind nebula (sometimes called a plerion, a diffuse nebula found inside the shells of supernova remnants, like what we see in the Crab Nebula); and mixed morphology ("thermal composite") where central thermal X-ray emission is seen, enclosed by a radio-bright shell, with examples of this class being SNR's W28 and W44. Supernova remnants are thought to be major sources of galactic cosmic rays, a connection first suggested by Walter Baade and Fritz Zwicky in 1934. Supernova remnants can produce the shock fronts that are energetic enough to generate very high energy cosmic rays, as has been seen in observations of SN 1006, which show synchrotron emission consistent with a cosmic ray source.

Table 3.3. Some challenging (and not so challenging) supernova remnants [59]

Name	Date of arrival of supernova's light at Earth	App. Mag.	Distance, LY	Type	Remnant
Sagittarius A East	–	–	26,000	–	Sagittarius A East
Simeis 147	>100,000 B.C.	–	3,000	–	Simeis 147 or sharpless 2-240
W49B	–	–	35,000	–	GRB Remnant?
W50	–	–	16,000	–	SS 433
Vela Supernova	11th–9th millennium B.C.	–	800	–	Vela supernova remnant
Veil Nebula	>3600 B.C.	–	1,400–2,600	–	NGC 6960 6974, 6979, 6992, and 6995
Puppis A	~1700 B.C.	–	7,000 approx	–	
SN 185	7 Dec. 185	–	8,200?	Ia?	Possibly RCW 86
SN 1006	1 May 1006	−7.5	7,200	Ia	SNR 1006
SN 1054	1054	−6	6,300	II	Crab Nebula
G350.1-0.3	About 1100	–	15,000 approx	–	G350.1-0.3
SN 1181	1181	−1	26,000 at least	–	Possibly 3C58
RX J0852.0-4622	About 1250	–	700	–	G266.2-12
SN 1572	11 Nov. 1572	−4	7,500	Ia	Tycho's supernova remnant
SN 1604	8 Oct. 1604	−2.5	20,000	Ia	Kepler's supernova remnant
Cassiopeia A	Mid-seventeenth century	+6	10,000	IIb	Cassiopeia A supernova remnant
G1.9 + 0.3	About 1868	–	25,000 approx.	–	Supernova remnant G1.9 + 0.3
SN 1885A	20 Aug. 1885	+6	2,500,000	–	SNR 1885A
SN 1987A	24 Feb. 1987	+3	168,000	II-P	SNR 1987A
E0102	–	–	190,000	–	E0102

Many of these supernovae remnants provide faint object challenges to rival those on lists presented later in this book. One such object listed in one of the "Off the Deep End" tables is Simeis 3-210, which is actually a thin filament that passes through a magnitude 6.4 star. There are fainter parts of the Veil Nebula that are challenging to observe, segments not often observed but can be tracked down by the persistent observer with dark skies and large aperture. Table 3.3 contains a list of some challenging and not-so-challenging supernovae remnants.

Accessibility of These Objects

Many of the objects presented in tabular form in Chap. 10 are in the form of clusters and nebulae. There are also several Astronomical League (AL) observing clubs dedicated to objects such as open clusters, globular clusters, and planetary nebulae. (These are not presented in Chap. 10 but are referenced in Chap. 9 and Appendix D of this book.) The tables provided in Chap. 10 contain a substantial sampling of HII (emission nebulae) regions, globular clusters in other galaxies, HII regions in other galaxies, old and large planetary nebulae, pairs and groups of galaxies, and so on.

Globular clusters can be seen in other galaxies, and the 75 brightest of M31 (the Andromeda Galaxy) are also presented. There are a handful of other galaxies within which globular clusters can be seen, such as NGC 1049, the brightest such object in the dwarf galaxy Fornax A. A plethora of other objects can be observed in M33 and M101; these mostly take on the form of knots and star clouds, but also visible are the brighter HII regions, stellar associations, and star clusters. NGC 604, of M33 (the Pinwheel Galaxy in Triangulum) is one of the largest emission nebulae known – more than 1,000 times bigger than the Orion Nebula and easily seen in medium-sized 'scopes under dark skies. Several other smaller nebulae are also accessible with large 'scopes, dark skies, and OIII or other nebular filters.

Most observationally accessible nebulae in the sky (that is, those you can see with a large backyard telescope and necessary accessories) are objects that exist in the Milky Way Galaxy. Large objects can be seen from great distances (assuming no dimming or extinction from the interstellar or intergalactic mediums), and this includes objects that reside in other galaxies, as mentioned earlier. This last situation enables one with a large telescope and an image or a map of the objects in other galaxies to find these objects visually for themselves. A sampling of such objects for M33 is provided later in this book.

Also included in the book are challenging planetary nebulae, such as the PK (Perek-Kohoutek) nebulae. These can be challenging due either to their small apparent size (stellar in appearance) or their low surface brightness (necessitating dark skies and/or special filters to make them visible). These challenging planetaries are also presented as observing targets in several tables in Chap. 10.

Chapter 4

The Nature of Galaxies and Galaxy Clusters

Introduction

Galaxies are huge collections of stars and vast amounts of gas and dust and are found throughout the known universe. Galaxies outside our own are so far away that the stars blend together into an amorphous haze, and some of the dust (in larger concentrations) is visible in silhouette against the bright material (e.g., the dust lane of the Andromeda and Sombrero galaxies). They come in various morphologies and are categorized into four groups based on their appearance in the sky: ellipticals, spirals, barred spirals, and irregulars. They can contain as few as tens of millions of stars but as many as hundreds of trillions of stars.

Galaxies were not noticed at first because of their faintness, but when telescopes became large enough to gather significant amounts of light, they began to come into focus [61]. In 1845, William Parsons, the third Earl of Ross, built a 72-in. telescope in Ireland. With it he discovered that some of the nebulae had spiral shapes (Fig. 4.1), and he concluded immediately that these were great spiral clouds of stars blurred together. Immanual Kant, a German philosopher, in 1755 coined the term "island universes" (great wheels of stars), which Parsons later adopted. Sir William Herschel counted stars in different directions and from these star counts came up with the description of the Milky Way's shape as resembling that of a huge grindstone. He thought that spiral nebulae were whirls of gas and faint stars within the Milky Way Galaxy.

At first, photography (when it began to be used for astronomical applications in the late 1800s) showed little detail in these objects. However, with the development of telescopes and new techniques, all that changed. With the 100-in. telescope that became operational in 1917, then the largest such instrument on Planet Earth, for the first time, as also noticed by Hubble, galaxies began to be resolved into individual stars. Hubble also noticed that some of these stars varied regularly. The light curves of these stars revealed them to be Cepheids, a special type of variable star whose period-luminosity relationship was established only years earlier (Henrietta Swan Leavitt established this relationship for classical Cepheids in 1908). In 1924, Hubble used this relationship between the pulsation period and the luminosity of the star to establish that these stars were well outside our own Milky Way Galaxy. These had become known as Cepheid variables since 1784, months after the prototype, delta Cephei, had been found to vary in this pattern. (Interestingly, eta

B. Cudnik, *Faint Objects and How to Observe Them*, Astronomers' Observing Guides, DOI 10.1007/978-1-4419-6757-2_4, © Springer Science+Business Media New York 2013

Fig. 4.1. The sketch made by Lord Rosse of the Whirlpool Galaxy in 1845 [60].

Aquilae was the first Cepheid to be discovered, but it was not the first in which the Cepheid pulsation pattern was established.) Hubble's work showed that the so-called "spiral nebulae" were actually whole other systems of stars outside our own system of stars that we call the Milky Way.

Microwave radiation of wavelength 21-com was predicted to exist as early as 1944, and this was actually observed in 1951 [62]. This radiation originates from interstellar atomic hydrogen gas, and it allowed for much improved study of the Milky Way Galaxy, since it is not affected by interstellar material and its Doppler shift can be used to map out the motion of the gas in the Milky Way Galaxy. Much of this gas, which lay along the spiral arms, revealed a bar structure in the center of the galaxy, rendering it a barred spiral and not an ordinary spiral, as has been assumed all along.

It was also discovered, in the 1970s, that the luminous mass does not properly match the speed profile of the rotating gas. Instead of the velocity slowing with distance from the center, like what we see in the Solar System, we observe a flat rotation velocity curve, almost as if the galaxy were a rigid rotating disk. This was observed in the Milky Way and in many other edge-on galaxies. The cause of this discrepancy was originally thought to be unseen dwarf stars and/or massive subatomic particles, but the improved observations brought by the Hubble Space Telescope established that these types of materials are insufficient to produce the observed effects. There must be some type of exotic or "dark" matter that only shows its presence through its gravitational influence.

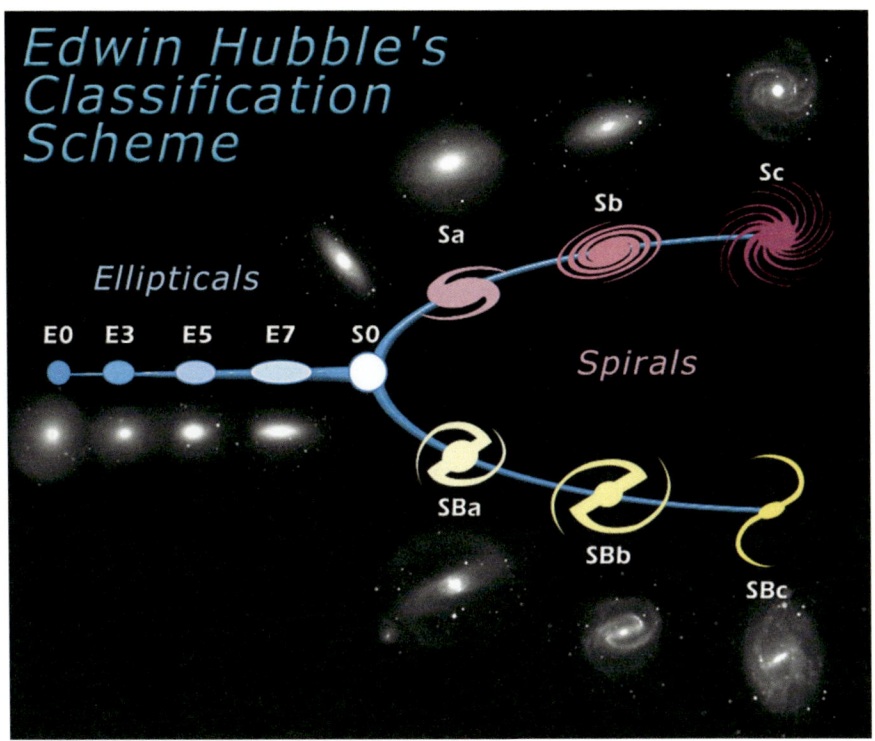

Fig. 4.2. Hubble tuning fork diagram (Courtesy of NASA and ESA).

Today, telescopes in space and on the ground are capable of detecting up to a few hundred billion galaxies, and more will likely be revealed with new telescopes. This number is based on estimates made as a result of the HST deep field exposures. Indeed, more than 99% of visible galaxies in the known universe can only been seen through such deep field surveys. Galaxies in general can be classified under three main types: elliptical, spiral, and irregular. Elliptical galaxies tend to be gas and dust poor, whereas spirals and barred spirals usually are gas and dust rich. Irregular galaxies range from gas/dust poor to gas/dust rich. This book will focus on observing the faint, nearby irregular galaxies and dwarf elliptical galaxies (these are quite close, galactic speaking, but intrinsically faint). The focus of the book will also be on larger but very distant faint galaxies and galaxy groups such as Stephan's Quintet and the Hickson group of galaxies.

Galaxies can be found in a wide variety of shapes and sizes. It was not until the twentieth century that the true nature of galaxies was appreciated. The astronomer Edwin Hubble was one of the individuals who were instrumental in finding this out, and he also constructed a scheme to classify these objects. They were difficult to classify at first because of their faintness in the sky, but with the advent of astro-photography (first by analog methods and later by digital) the scheme of classification became possible. The most widely used scheme is the Hubble tuning fork (Fig. 4.2), which attempts to classify galaxies based on trends in their physical appearance.

Hubble noticed two main groups of galaxies, the spiral galaxies and the elliptical galaxies. He also noticed that some of the spirals had a straight, bright line, or bar, running through their centers and classified these as "barred spirals." Hubble's tuning fork was thought by Hubble himself to describe the evolution of galaxies over time, with galaxies starting as a round elliptical; then, over time, the galaxies would flatten and then develop spiral arms, which would start out tight but loosen over time. Looking at the pattern shown by the Hubble tuning fork it is easy to see that it would make sense for it to be an evolutionary diagram, since the morphologies seem to "flow" from one end to the other. However, research conducted since the time of Edwin Hubble has shown that galaxies do not evolve as Hubble thought, but they actually form as a result of mergers and acquisitions of smaller objects by larger ones [63].

The following are brief descriptions of the main classifications of galaxies, along with a comparison of each type to our home galaxy, the Milky Way.

Spiral Galaxies

These vast systems of stars, gas, and dust are disk-shaped, have spiral arms, and many of these have a bar-shaped nucleus or inner region [64]. In fact, about two-thirds of all spirals are known to be barred spirals. These galaxies tend to be rich in gas and dust and have significant star formation activity, as displayed by emission nebulae and an abundance of hot, bright stars.

Spiral galaxies (Fig. 4.3) are subdivided based on the size of the nucleus, the gas and dust content, and how tightly wound the spiral arms are. Sa-type galaxies tend to have larger nuclei; less gas and dust; fewer hot, bright stars; and tightly wound spiral arms. Sc-type galaxies have small nuclei, lots of gas and dust, many bright stars, and loose spiral arms. Sb-type galaxies are intermediate between Sa and Sc, and Sd-types are even more extreme than Sc.

Once galaxies were discovered to be entities outside of the Milky Way Galaxy, the challenge was to pin down fairly accurate distance estimates of these objects. Cepheid variable stars are only useful out to about 100 million light years (30 Mpc), so other methods, such as using supernovae, need to be applied. Distance is the first of three fundamental parameters that astronomers are interested in; the other two are size (including diameter and mass) and luminosity.

The cosmological distance ladder is a hierarchy of methods that astronomers use to figure out how far away things are in space. The most fundamental of these is a method called trigonometric or stellar parallax, where the distances to stars are directly determined by triangulation. The positions of stars are measured from both ends of Earth's orbit and the distance is reckoned from there. The HIPPARCOS (*HI* Precision *PA*Rallax *CO*llecting Satellite) has measured, during its 3.5-year mission in the early 1990s, the parallax of 118,218 stars with the highest precision, and 1,058,322 stars with lesser precision. Further work resulting in the Tycho 2 catalog, completed in 2000, brings the total to 2,539,913 stars [65].

The parallax method is used for subsequent methods to gauge the distances to more and more distant objects. Cluster main-sequence fitting is used to determine not only the age of a star cluster but also its distance, by comparing the apparent brightness of its member stars with the absolute brightness of stars of their spectral type. Cepheid and RR Lyrae (the fainter cousins of the Cepheids that operate

Fig. 4.3. This is a typical spiral galaxy, NGC 4414 in the constellation Coma Berenices. The object is about 55,000 light years (16,900 pc) across and approximately 60 million light years (18.5 million pc) away from Earth (This image is courtesy of NASA and ESA and was created for NASA by Space Telescope Science Institute).

on the same principle) variable stars use a relationship between the period of pulsation and the luminosity of the star to find its absolute luminosity and get its distance. Cepheids have been used in places as far away as the Virgo Cluster of galaxies, some 65 million light years distant.

Next are the supernovae, which are among the brightest objects in the universe. Although all types are at least somewhat useful, the Type I supernovae are most useful in that they are caused by the nuclear detonation of a carbon white dwarf reaching a mass known as the Chandrasekhar limit (1.46 solar masses). It does this by pulling material from a nearby companion star. Under the right conditions these stars go "over the edge" and explode, and since they all start with the same mass, so it is thought, they should all reach the same peak brightness and indeed have the same, or similar, shaped light curves. These enable distance measurements out to several hundreds of millions of light years.

Back to galaxies. The Milky Way Galaxy is larger and more luminous than most spirals, but there are some that are larger still. The largest known spiral galaxy has nearly four times the diameter, twice the mass, and ten times the luminosity of our home galaxy.

Some of the pressing questions in galactic astronomy are (1) Where do spiral galaxies get their spirals, and (2) Once they get their spirals, how do they keep them? [66]. And why are these galaxies spiral-shaped to begin with? Consider the

Fig. 4.4. The two faces of the Whirlpool Galaxy. The image at *left* is a visible light image showing the complex structure, including pink emission nebulae, clouds of stars, and the nucleus. The image on the *right* shows the dust structure of the galaxy as seen in the near infrared. Astronomers combined two images, one taken in the visible and one in near-infrared and subtracted the total amount of starlight from both images to see the galaxy's dust structure (in *red*). The smallest details are 35 light years across (Image courtesy of NASA and hubblesite.org).

example of M 51, the Whirlpool Galaxy (Figs. 4.4 and 4.5). Observations of this object by the Hubble Space Telescope hint at the presence of acoustic pressure waves rather than gravity density waves operating in the innermost 1,000 light year radius.

Galaxies were not even recognized as their true selves until the 1920s. After our view of them became clear by mid-twentieth century, various explanations have been put forth to explain what people were seeing. Spiral arms were thought to be density waves (1960s, C.C. Lin and Frank Shu of MIT) or traveling zones of compression and rarefaction (spreading out). These zones move independently of the matter in them, much like water waves, traffic jams, and sound waves move independent of the individual components (water molecules, vehicles, air molecules). Density waves take the shape of spirals, rotating independent of the material within and driven by gravity. The gravity of the waves holds higher concentrations of stars in place, making them visible in images of galaxies.

Density waves also suck in and squeeze clouds in the interstellar medium. This triggers star formation, which helps to populate the spiral arms. The process is analogous to the white caps or bubbles on the crests of water waves. The visible part of the spiral looks messier than the underlying wave due to the former being more fragile in the galaxy's turbulence.

But…where do these waves come from in the first place? Do they last a while or are they short-lived? Another analogy comes to mind here: a boat traveling in water, leaving a wake. Start the boat rotating in a little circle, so the waves spiral outward, and see what happens. Spiral patterns should be long-lived, because the

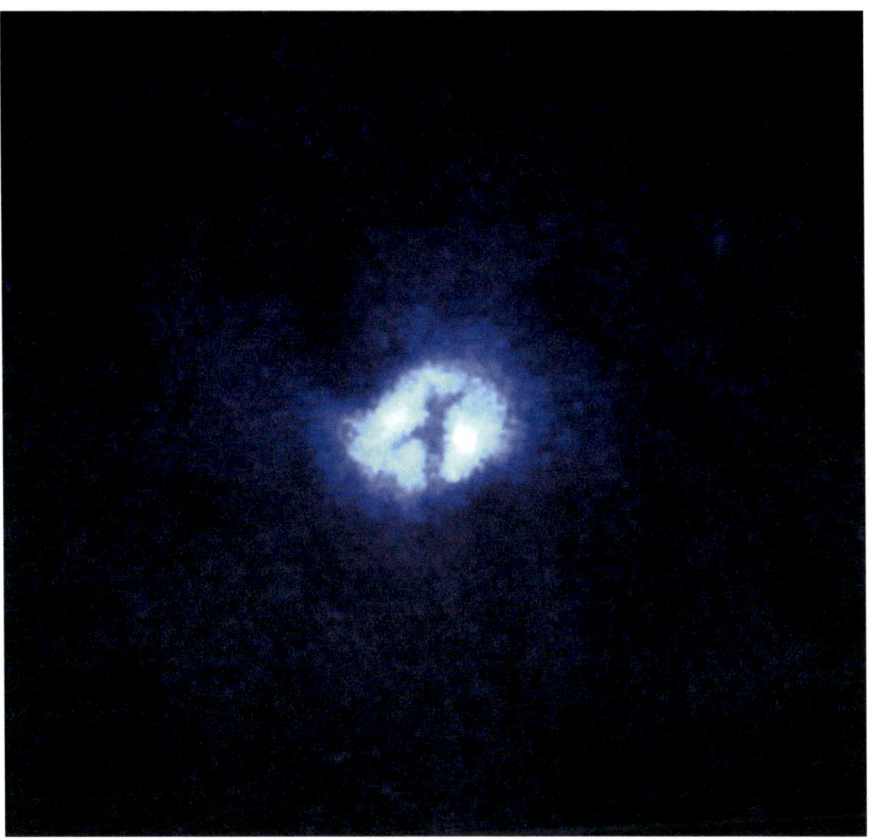

Fig. 4.5. The nucleus of the Whirlpool Galaxy (M51) showing the cross that marks the location of the central black hole. It was first thought that this feature was formed by two dust rings, but later interpretations indicate these are foreground features silhouetted by the active nucleus (Image courtesy of NASA and ESA).

pattern contains what are called standing waves, patterns of seemingly still waves that arise from the interaction between waves reflected off of a barrier and those approaching a barrier. These interact to build up the crests and deepen the troughs of a seemingly motionless wave.

Astronomer Dr. Toomre of MIT felt that most spirals looked messier (disorganized, not related to Charles Messier) and less symmetric than true standing wave systems (these are the "grand design" spirals), which are not expected to survive for long in galaxies. If we take a large lump of material orbiting a galaxy's core and observe what happens to it over time, it is stretched out by the tidal forces of the galaxy and is torn apart. Add a few more lumps and one is able to explain the "bumpy" texture that most galaxies show in images. Spiral arms change constantly with new spirals replacing old, dying spirals in a cycle that maintains the overall appearance of the galaxy. The changes can be dramatic over the course of 250 million years. But astronomers Lin and Shu say that these patterns may be stable for 2.5 billion years (as is the case of the Milky Way) or even longer.

Galaxy formation theory states that a small disturbance in a rotating galactic disk can launch spiral waves. The tug can also come from outside the galaxy: nearby spiral galaxies' tidal effects on bars in barred spiral galaxies may exert torques that get the waves started. Disturbances that are small to begin with can quickly grow to large scale visible patterns, but it is unclear as to what the driving mechanism is or what causes the pattern to last. Naval research scientist Xiaolei Zhang thinks stars may lose angular momentum and spiral inward. As they do so, they give their angular momentum to the spiral waves, maintaining the wave and causing the star to slowly spiral inward. Over the long term, this action turns a spiral's inner disk into a lump that looks more like an elliptical.

The above theory explains survey results from the Hubble Space Telescope. More galaxies are elliptical today than in the distant past, when spirals and ellipticals came in roughly equal numbers. The spiral signature, according to Zhang, is an expression of the system tending toward disorder, even though it assumes the appearance of order. Another analogy that is useful here is to our atmosphere – the sharp temperature difference between upper and lower parts of it. This causes organized patterns of convection air currents that work towards reducing the temperature difference.

Sellwood's angular momentum gap theory states that stars' angular momentum distributions in galaxies may produce one or more gaps that lead to the formation of spiral arms.

Like the Milky Way Galaxy, spiral galaxies house large numbers of deep sky objects. Open clusters are largely found in the spiral arms, where gas densities are highest, providing raw material for star formation. Open clusters are strongly concentrated close to the galactic plane, as open clusters usually disperse before they had much time to travel far above or below the plane of the galaxy or far from a spiral arm.

Elliptical Galaxies

These galaxies have no disk or spiral arms and almost no gas and dust [64, 67]. They range from small dwarfs to huge giants and have a more featureless and redder appearance than spiral galaxies, due to the dominance of old, red stars. Their three-dimensional shapes are not certain, but the more elongated specimens may be shaped like flattened spheres or American footballs or flattened loaves of bread. These are sub-classified according to their shape: E0 is round, E7 (the highest it goes) is highly elliptical.

Elliptical galaxies come in a wide range of sizes. The largest ones are called giant ellipticals (Such as the one featured in Fig. 4.6) and can be as large as five times the diameter of the Milky Way, 50 times the mass, and five times the total luminosity. Many ellipticals are small objects called dwarf ellipticals, with masses as small as 0.01% the mass of the Milky Way, 1% its diameter, and 0.005% of its luminosity. Dwarf elliptical galaxies may be no larger than the average Milky Way globular cluster, but they do have something globulars lack – dark matter [68].

Open star clusters are not seen in elliptical galaxies due to their gas-poor natures. Star formation in an elliptical ceased a long time ago, so that any open cluster that was present had long since dispersed. These galaxies have sparse interstellar media with very little gas or dust between the stars, resulting in few open star clusters and few large stars. The dynamical properties of elliptical galaxies and the bulges of disk galaxies are similar, which suggests they form from similar

Fig. 4.6. A cluster of diverse galaxies, Abell S0740, as seen by Hubble. The image is dominated by a giant elliptical galaxy ESO 325-G004, which has 100 billion Sun's worth of mass and is over 450 million light years away in the direction of the constellation Centaurus. There are thousands of globular star clusters orbiting ESO 325-G004, which appear as pinpoints of light within the diffuse halo (Image courtesy of NASA and hubblesite.org).

physical processes (but this is still not certain). The motions of stars in both entities are predominantly radial (vs. that in the disks of spiral galaxies, which is dominated by rotation).

These galaxies tend to be easier to see through backyard 'scopes because of their relatively high surface brightness. Yet they generally appear as featureless balls of haze, some round, some flattened, but most with bright cores. However many elliptical galaxies lack the bright core that would make them easy to see; they are much harder to see because of their low surface brightness. Some of these galaxies are close enough to enable their attendant globular clusters to be seen, but few, if any, ellipticals that display emission nebulae or planetary nebulae are visible from ground-based telescopes used by amateurs (Fig. 4.7 shows two examples near both ends of the elliptical galaxy size spectrum).

Fig. 4.7. A few examples of elliptical galaxies: (**a**) giant elliptical M87 and several others in the heart of the Virgo Supercluster (Image courtesy of Paul and Liz Downing), (**b**) dwarf spheroidal Leo I (Image courtesy of Loyd Overcash).

Miscellaneous Types of Galaxies

Lenticular Galaxies

The Lenticular, or S0 (and SB0), galaxies are more regular in shape. They have a disk and an obvious bulge, but no visible gas and dust, no obvious spiral arms, and few to no hot, bright stars. These are called "Lenticular" galaxies because of their resemblance (especially when viewed nearly edge-on) to convex lenses. They

Fig. 4.8. NGC 5866, the Spindle Galaxy, an example of an edge-on lenticular galaxy (Image courtesy of NASA, ESA, and Space Telescope Science Institute).

are the intermediate types of galaxies in terms of morphology, between elliptical and spirals or barred spirals. These galaxies have used up or lost most of the raw material that is needed to make stars, which leads to their rather bland appearance. There are barred lenticular galaxies that display a central bar and are classified as SB0.

Some lenticular galaxies do have dust absorption, like the Spindle Galaxy (NGC 5866, Fig. 4.8) and NGC 2787 (Fig. 4.9). The former is an example of an edge-on lenticular, and the latter is more face-on. Aside from the dusty arcs, it is difficult to tell face-on lenticulars from elliptical. In fact both share common properties, such as spectral features and stellar types that dominate the population of the galaxy (mostly older Population II stars).

Fig. 4.9. NGC 2787 (Image courtesy of NASA/ESA and the Hubble Heritage Team, STScI/AURA).

Irregular Galaxies

Irregular galaxies do not have a regular shape, as their names imply, but they tend to be rich in gas and dust. These are chaotic mixtures of gas, dust, and stars, with no obvious shape or structure. The Magellanic Clouds that are two of over a dozen galactic satellites of the Milky Way are bright examples of irregular galaxies that grace our Southern Hemisphere skies. They are sites of active star formation and are chock full of emission nebulae, such as the Tarantula Nebula.

Open star clusters may be found throughout the galaxy, but their highest concentration is where the density of the gas is highest. Again, this is where the raw material for star formation is found.

Irregular galaxies (Fig. 4.10) are small, with between 1% and 25% of the mass of the Milky Way. In fact, ellipticals have properties that range in the following manner: mass is from less than 0. 05% up to 15% of the mass of the Milky Way; diameters range from 5% to 25% that of the Milky Way; and masses are anywhere from 0.005% to 10% that of our home galaxy.

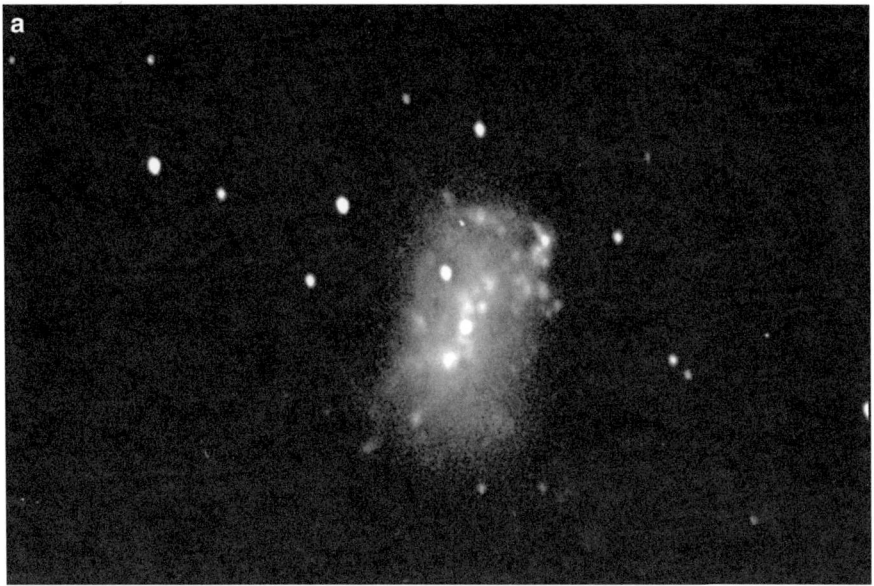

Fig. 4.10. Some examples of irregular galaxies: (**a**) NGC 4449 (Image courtesy of Paul and Liz Downing), and (**b**) IC 4662 in great detail (Image credit: NASA, ESA, and K. McQuinn, University of Minnesota, Minneapolis).

There are two broad types of irregular galaxies. Irr-I objects have some structure but do not align neatly with the Hubble tuning fork classification scheme. Irr-II galaxies have no structure that would resemble a Hubble classification scheme and may have been disrupted at some point in their past.

Ring and Starburst Galaxies

Galaxies can display an amazing variety of shapes and forms, from ring galaxies to galaxies resembling tadpoles. Peculiar galaxies are those that develop unusual shapes or parts due to tidal interactions with other galaxies. One example of a peculiar galaxy that may have had such interaction is a ring galaxy, an object that shows a ring-like structure of stars and materials surrounding a bare core. Ring galaxies are left behind after high speed collisions. They occur when one galaxy passes through another, and computer simulations seem to verify this. The Andromeda Galaxy features two dusty rings, a small inner ring and a larger outer ring. This was produced when M32 plunged through the disk of M31.

Other types of collisions, such as a glancing "blow" can produce tidal tails and other interesting features. These shapes of peculiar galaxies can only be faintly made out visually through the larger backyard telescopes, but there may be a few exceptions to this. There is a list of such galaxies featured in Chap. 10, and these are interacting galaxies (two examples are the Mice and the Antennae; others are shown in Fig. 4.11).

Collisions and interactions may trigger explosive star formation in galaxies, which are given the term "starburst galaxy." A starburst galaxy is characterized by

Fig. 4.10. (continued)

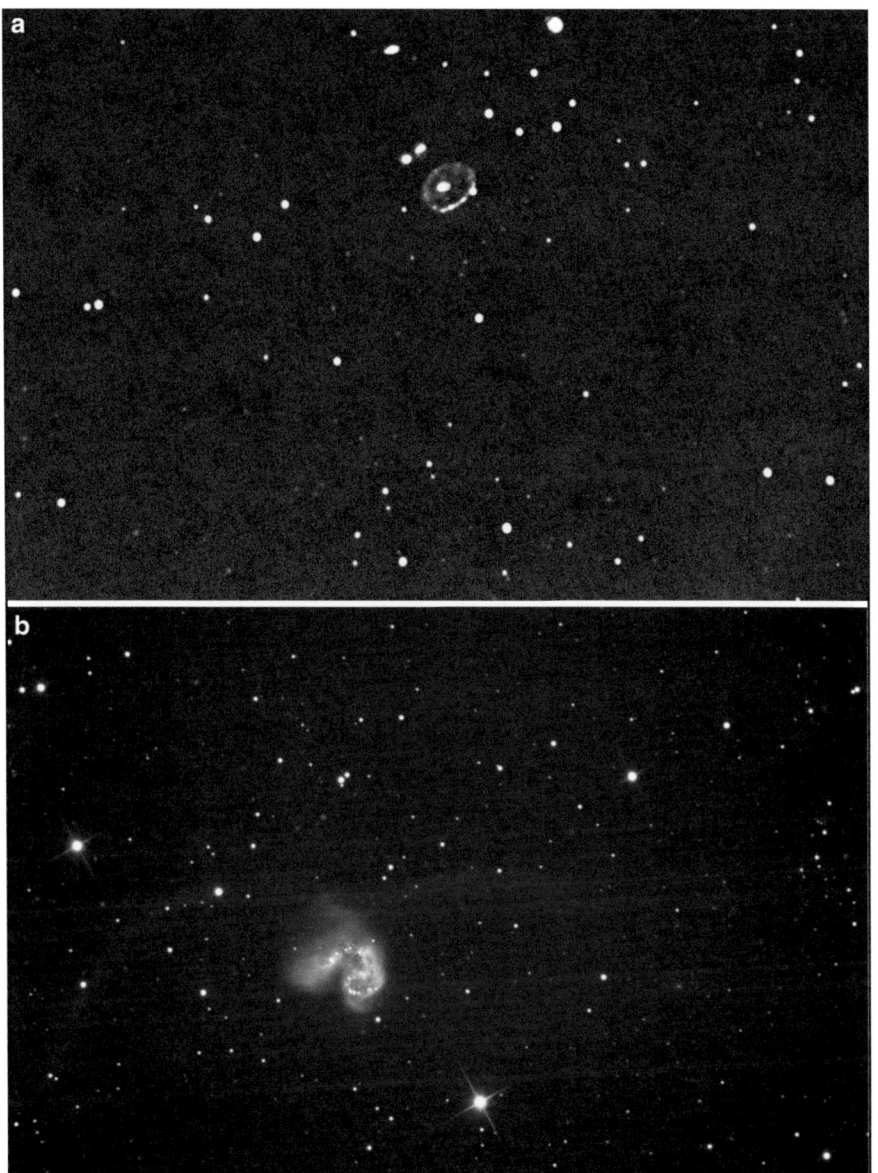

Fig. 4.11. Examples of interacting galaxies: (**a**) the Cartwheel Galaxy (PGC 2248, courtesy of Paul and Liz Downing), (**b**) NGC 4038/9, "the Antennae" (Image courtesy of Loyd Overcash), and (**c**) NGC 4567/8, the "Siamese Twins" interacting galaxy pair (Image courtesy of Loyd Overcash).

dusty concentrations of gas and the appearance of newly formed stars, to include massive, bright stars that ionize the surrounding clouds of hydrogen gas to produce H-II regions. These massive stars live only a short time, cosmologically speaking, and produce supernova explosions that push remnants outward, causing them to interact vigorously with the surrounding gas. A chain reaction is set up that goes like fireworks as stars are built in an expanding regions that spreads throughout

Fig. 4.11. (continued)

the gaseous region. Only when these raw materials (cold gas, giant molecular clouds) are almost completely consumed or dispersed does the starburst activity come to an end. The galaxy M82 is a proto-typical starburst galaxy, made so after a close encounter with the larger M81 galaxy.

Dwarf Galaxies and the Local Group

Surveys of the Local Group of galaxies turn up two large galaxies: the Milky Way Galaxy and the Andromeda Galaxy (M31), a smaller spiral galaxy called the Triangulum Galaxy (M33), and fifty-one members that are dwarf galaxies [69]. The Milky Way's satellite system contains 14 dwarf galaxies, and Andromeda's satellite system has 24 dwarf elliptical, irregular, and spheroidal galaxies. The Local Group is situated on the outskirts of the Virgo Supercluster of galaxies. The vast majority of the nearest galaxies are dwarf galaxies, and by extension, these are likely the most common types of galaxies in the universe. They cannot be seen much beyond the Local Group or five million light years (1.6 million pc) distant.

Typical dwarf galaxies are about 1% the size of the Milky Way and contain only a few billion stars. Ultra-compact dwarf galaxies that are only about 100 pc (330 light years) across have recently been found. Although the Milky Way has some 14 satellite galaxies, some estimate that there are between 300 and 500 dwarf galaxies in the Milky Way's region alone. Dwarf galaxies may also be classified as elliptical, spiral, or irregular, and dwarf ellipticals are often termed dwarf spheroidal galaxies, since they bear little resemblance to large elliptical.

In the next section, we will look at interacting galaxies more closely.

Interacting Galaxies, Clusters, and Superclusters

Interacting Galaxies

Many galaxies come in pairs or threes (or more), and they can be seen interacting in one way or another with each other [70]. They distort each other gravitationally, producing tidal tails and star shells, and as a result, this action may trigger the formation of spiral arms. Large galaxies can disturb small galaxies, even cannibalizing them.

Some effects and after-effects of galactic interactions include episodes of explosive star formation, distortions of spiral shapes, formation of spiral arms, and features that act as evidence of past interactions and mergers. Observational evidence of such interactions in the past includes features such as star streams and multiple nuclei. In the Milky Way Galaxy, there is ample observational evidence of past events, including the pulling apart of the Sagittarius dwarf galaxy. Astronomers also observe great streams of stars wrapped around the galaxy as a result of the absorption of the Canis Major dwarf galaxy.

Interactions between galaxies are frequent, since the separation between galaxies within a cluster, for example, averages about 10 or 12 times their diameter. These events play a role in the evolution of galaxies. Near misses result in warping distortions in each galaxy due to tidal interactions, and collisions occur when one passes through another. Since the spacing of stars in each galaxy is much larger than the sizes of the stars themselves, they pass right by each other without themselves colliding (since the stars are essentially point sources within the galaxies). Clouds of gas and dust, occupying much larger volumes of space, do collide with each other and may trigger bursts of star formation as the interstellar media of each galaxy becomes compressed and disrupted.

In other cases, two or more galaxies may merge. This is unlike collisions, where each galaxy essentially keeps going after the collision, since they have more than enough relative momentum not to merge. Mergers happen when two or more galaxies do not have enough momentum to just "keep going" and so they fall into themselves. Mergers can result in significant changes to the morphology of either of the individual galaxies after an encounter versus the morphologies of the original galaxies. The final product of the two galaxies may be of a different type altogether. If one galaxy is much larger and more massive than the other, then the result is termed "cannibalism," where the smaller of the two is torn apart and absorbed while the larger of the two is little changed. The Milky Way Galaxy is in the process of cannibalizing several galaxies as evidenced by various faint star streams visible in some locations. The "victims" (which have already been named) include the Sagittarius dwarf elliptical galaxy and the Canis Major dwarf galaxy.

Computer simulations of galaxy collisions are often fun to watch, not to mention scientifically useful and insightful. The simulations take one or two billion years or so and compress this time into a few minutes. That's because galaxies are so big, and the events play out so slowly compared to what we are used to as far as time, that we can only see snapshots of different types of collisions in various stages all over the universe. The simulations generally simulate several thousand stars per galaxy and calculate the motions of each point or ensemble of points as the event plays out. Snapshots of computer simulations closely match images of real-life galaxy collisions.

Clusters of Galaxies

Most galaxies are found in groups ranging from a few individuals to a few thousand galaxies occupying volumes from 1 to 10 Megaparsecs across [71]. The Milky Way is a member of a cluster of nearly 40 galaxies, and this is but one of thousands of other clusters of galaxies that we know of. Galaxy clusters can be divided roughly into two groups: rich clusters and poor clusters.

Clusters are larger than groups, but there is no sharp distinction between the two, and they are held together by their mutual gravitational attraction [72]. Upon closer inspection of galaxy clusters, astronomers noticed some unusual characteristics of the clusters. The velocities of the member galaxies seem rather large for them to be bound in the cluster, and this implied a lot more mass than what was directly observed. Also, observations of these objects in the X-ray part of the spectrum reveal large amounts of intergalactic gas referred to as the intracluster medium. This gas is very hot, on the order of 10–100 million K and emits X-rays in the form of bremsstrahlung and atomic line emission.

The total mass of the system, noting that the intracluster medium gas is in hydrostatic equilibrium and the galaxies do not appear to be escaping from the cluster, is approximately six time larger than the mass of the galaxies or the hot gas. This missing component is dark matter whose physical nature is unknown and makes up 85% of the mass of a typical cluster (the galaxies make up 5% of the mass and the hot X-ray emitting gas makes up 10%). Observations of gravitational lensing associated with the Bullet Cluster (Fig. 4.12), which consists of two colliding clusters of galaxies, is claimed to be the best evidence of the existence of dark matter.

There are alternative interpretations of phenomena attributed to dark matter [73], which attributes the effects observed to unseen ordinary matter (or non-luminous baryonic matter) rather than cold dark matter. This, combined with Modified Newtonian Dynamics, is said to adequately explain the behavior with this and other galaxy clusters. Modified Newtonian Dynamics (MOND), proposed by Mordehai Milgrom in 1983, states that acceleration due to gravitational force is not linearly proportional to force at small values.

Rich Clusters of Galaxies (Superclusters)

These clusters contain one thousand or more individual galaxies, many of which are elliptical. In fact, 80–90% of their members are elliptical and lenticulars, with a few spirals. These rich clusters of galaxies almost always have a central concentration, with one or more giant ellipticals dominating the region. The nearest example of one such cluster is the Virgo Cluster, whose center is about 17 Mpc (Mega – or million pc, equivalently 55 million light years) away from us and contains more than 2,500 galaxies. The central region is dominated by the giant elliptical M87. A second giant elliptical, M86, lies nearby.

Many rich clusters of galaxies, also known as superclusters, are filled with a hot gas that acts as an intracluster medium. The gas comes from mass leftover from

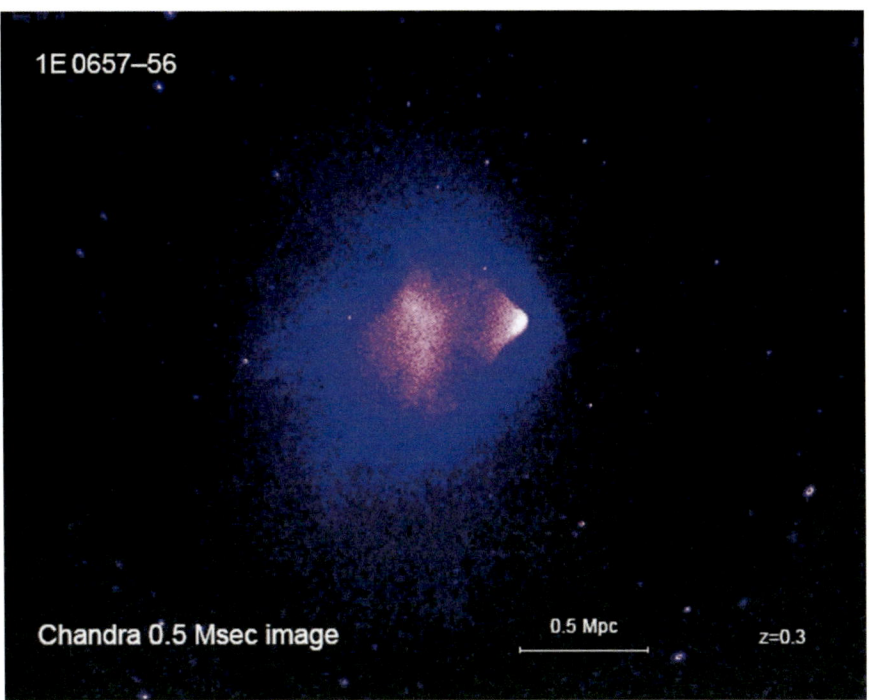

1E 0657–56

Chandra 0.5 Msec image

0.5 Mpc

z=0.3

Fig. 4.12. X-ray image from the Chandra X-ray observatory showing the Bullet Cluster (1E0657-56) (Image courtesy of NASA).

galaxy formation but some of the gas was driven out by supernova explosions from individual member galaxies. Also permeating the cluster is dark matter, which shows itself by gravitational lensing. This works as follows – more distant objects behind the galaxy cluster are made visible by the lensing effect of the overall gravitational field of the cluster. These more distant objects appear as distorted and curved images of galaxies scattered among the cluster members. Another manifestation not as obvious is the presence of hot intercluster gas (which shows itself by X-rays); the amount observed is only possible if the gas was contained by dark matter.

There are many examples of rich galaxy clusters whose member galaxies show up in the image along with curving, distorted images of more distant galaxies (Fig. 4.13). The gravitational field produced by the dark matter permeating the cluster changes the direction of the light coming from the more distant galaxies, acting like a glass lens and focusing the light in our direction. This results in more distant galaxies appearing distorted, but brighter than what they would otherwise appear from Earth.

Poor Clusters of Galaxies (Known Simply as Clusters)

These clusters of galaxies contain fewer than one thousand galaxies, and often they may contain only a few galaxies. These objects are scattered over a volume that is as large as a rich cluster, meaning that individual galaxies are widely separated.

Fig. 4.13. Abell 370 galaxy cluster imaged by Hubble Space Telescope and showing detailed images of gravitational lensing effects (Image Credit: NASA, ESA, the Hubble SM4 ERO Team, and ST-ECF).

The poor clusters have a larger percentage of spiral galaxies than rich clusters and lack a central concentration of galaxies. Clusters tend to have sub-clusters, with smaller galaxies crowded around larger ones. An example close to home is the so-called "Local Group" of galaxies that occupy a roughly spherical volume of space roughly 1 Megaparsec in diameter. The Local Group contains, as its brightest 32 members: 15 elliptical galaxies, four spirals, and 13 irregulars.

By the way, it is quite rare to have galaxies by themselves, but this does occur. Of these rarities, between 80% and 90% are spiral galaxies.

Conclusion

Just like the Milky Way Galaxy, all galaxies are made up of a mixture of stars and gas. Some stars have higher proportions of gas than others. We see galaxies across huge distances because of the combined light output of the brightest stellar minority (a small fraction of the total number of stars that gives off most of the total light of the system). In fact, the Sun would be well hidden from view to an observer looking at the Milky Way from the outside.

At the heart of the Milky Way and most larger galaxies is a supermassive black hole, ranging in mass from a million to a few billion suns' of material [71]. The size of the central bulge size is related to black hole size: the bigger the central bulge, the bigger the central black hole. The Milky Way Galaxy hosts a black hole with four million solar masses' worth of material. In spite of these huge sizes, central black holes make up only a tiny fraction of the total mass of a galaxy, from 1% to 0.001%.

Dark matter is a common feature in at least the larger galaxies (including the Milky Way), and it reveals itself in the galaxies' rotation curves. What this means is the following: the measured rotation rates of the various parts of the galaxy do not match those expected by the visible part alone based on Kepler's laws of orbital motion. There has to be some extra material "speeding things along" as one moves outward from the center of the galaxy. This makes the rotation of the galaxy resemble a rigid disk more than a fluid disk. This was one of the many observations that led to the postulation of the presence of dark matter. It is estimated that 90% of all matter in the universe is dark matter. Dark matter also permeates the clusters of galaxies.

Galaxies are the most common of the faint objects, literally numbering in the hundreds of billions (at least). The observable universe is estimated to contain at least several hundred billion galaxies, but the vast majority of these are not visually accessible with backyard telescopes (no matter how large the telescopes are). However there are at least several hundred thousand galaxies that can be seen through the largest backyard telescopes, and this book provides a small sample of these as part of the lists given in Chap. 10. Most of the galaxies are faint due to distance, but there are a number of relatively nearby galaxies that are classified as dwarf galaxies that are difficult to observe because of their low surface brightness. There are also some relatively nearby galaxies that are viewed through the dust of the Milky Way galactic plane. This dust (along with the presence of many foreground stars) works to dim the more distant galaxies by up to several magnitudes. Indeed there are many, many external galaxies rendered invisible by Milky Way material.

.

Chapter 5

The Nature of Quasars and Other Exotics

Introduction

In addition to the "normal" deep sky objects featured in the observing lists offered in Chap. 10, some of the brighter members of exotic objects, such as blazars, quasars, BL Lacertae (Lac) objects and others are also featured [74]. These include a handful of examples of gravitationally lensed objects such as the Ursa Major double quasar and "Einstein's Cross." An active galactic nucleus (AGN) is a compact region at the center of a galaxy that has a much higher than normal luminosity over part or all of the electromagnetic spectrum. A galaxy that hosts an AGN is termed an active galaxy. The radiation that comes from an AGN is thought to result from the accretion of mass by the central supermassive black hole.

Table 5.1 provides an overview of the properties of normal galaxies and the various kinds of active galaxies. Of these objects, the normal, starburst, and the two types of Seyferts are not radio loud, the quasar is 10% radio loud, and the blazar, BL Lac object, Optically Violent Variable (OVV) and radio galaxy are each radio loud.

Active Galactic Nuclei/Seyfert Galaxies

General Properties

What follows is a brief description of some of the important features of AGN. There is no single hallmark observational feature that defines an AGN but rather there are several types of observational characteristics that are found in a given AGN.

- NUCLEAR OPTICAL CONTINUUM EMISSION: Visible when we have a direct view of the central accretion disk, and jets can contribute to this emission. This has a roughly power-law dependence on wavelength.
- NUCLEAR INFRARED EMISSION: Visible whenever the accretion disk and its surroundings are obscured by gas and dust close to the nucleus and re-emitted; it is thermal emission that can be distinguished from any jet or disk-related component.

B. Cudnik, *Faint Objects and How to Observe Them*, Astronomers' Observing Guides,
DOI 10.1007/978-1-4419-6757-2_5, © Springer Science+Business Media New York 2013

Table 5.1. Overview of various kinds of AGN as compared with normal and starburst galaxies (Courtesy of Wikipedia)

Type	AGN	Narrow emission	Broad emission	X-ray	UV excess	Far-IR excess	Strong radio	Jets	Variable
Normal	No	Weak	None	Weak	None	None	None	None	No
Starburst	No	yes	No	Some	No	Yes	Some	No	No
Seyfert I	Yes	yes	Yes	Some	Some	Yes	Few	No	Yes
Seyfert II	Yes	yes	No	Some	Some	Yes	Few	Yes	Yes
Quasar	Yes	yes	Yes	Some	Yes	Yes	Some	Some	Yes
Blazar	Yes	No	Some	Yes	Yes	No	Yes	Yes	Yes
BL Lac	Yes	No	None/faint	Yes	Yes	No	Yes	Yes	Yes
OVV	Yes	No	Stronger than BL Lac	Yes	Yes	No	Yes	Yes	Yes
Radio galaxy	Yes	Some	Some	Some	Some	Yes	Yes	Yes	Yes

- BROAD OPTICAL EMISSION LINES: These come from cold material close to the black hole; the emission lines are broad because the emitting material is revolving around the black hole at great speeds, producing Doppler broadened lines.
- NARROW OPTICAL EMISSION LINES: These come from distant cold material that is revolving more slowly, hence the narrower lines.
- RADIO CONTINUUM EMISSION: Always due to a jet and whose spectrum is characteristic of synchrotron radiation.
- X-RAY CONTINUUM EMISSION: Can arise from both a jet and hot corona of the accretion disk via scattering; shows a power-law spectrum in both cases. "Soft excess" (the origin of this is unclear) in the X-ray emission present on top of the power-law component.
- X-RAY LINE EMISSION: Results from illumination of cold heavy elements by X-ray continuum. Fluorescence produces various emission lines, the best known of which is the iron line around 6.4 keV. Lines may be narrow or broad and point to the physical nature of the central black hole.

The type of active galaxy that one sees depends on the angle of perspective. The unified model of an AGN features a thick torus of material surrounding an accretion disk that surrounds a supermassive black hole. An AGN is what we see when we look straight down (perpendicular to the plane of the accretion disk) onto the black hole. A Seyfert galaxy is seen when we look at the black hole from an angle, so that we do not get the radiation directly (the radiation is beamed along an axis perpendicular to the accretion disk). A normal galaxy is observed when we are looking into the plane of the disk and torus, such that the torus obscures the active part from view.

More About AGN

Active galactic nuclei (AGN) were first observed in the 1950s as strong radio sources [75]. Improved technology in the 1970s showed that these galaxies shone brightly in many other wavelengths. Only a few percent of all galaxies in the universe are active. AGN tend to be both compact objects and extremely luminous on a persistent basis. The core object in all cases is a supermassive black hole (which

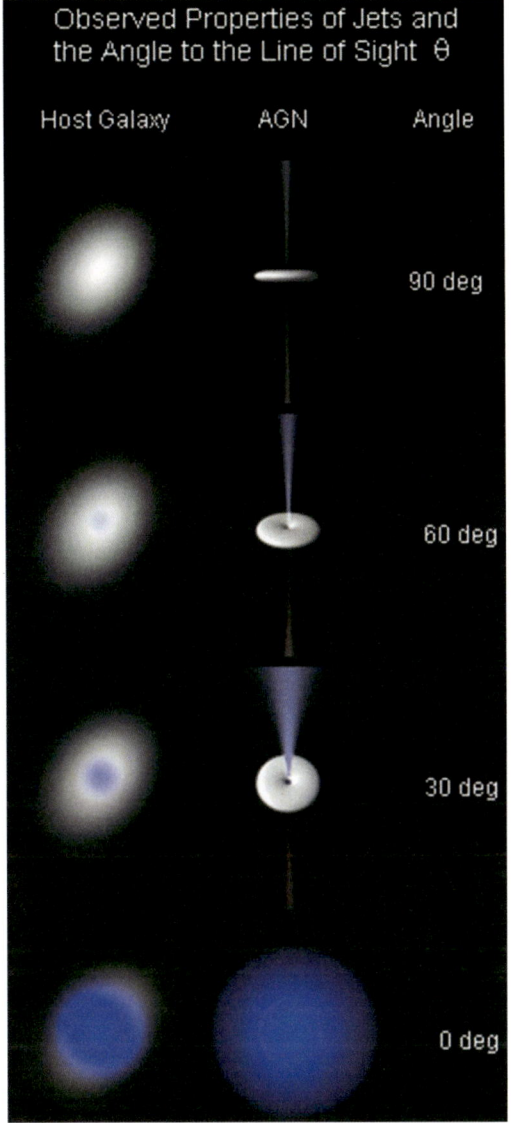

Fig. 5.1. Illustration of the unified model of AGN showing that the orientation of the jet with respect to our line of site determines what type of AGN we see (Image courtesy of Ron Kollgaard under Creative Commons Attribution).

is likely to exist in most or all massive galaxies). The common characteristics of AGN are that this black hole is consuming material.

Scientists, to explain the different and varied types of AGN as outlined in Table 5.1, use a unified model (an example of which is shown in Fig. 5.1) that consists of an accretion disk, relativistic jets, and a radiantly inefficient AGN. Unified models use a single type of object and different viewing perspectives to describe a particular AGN type. The currently favored unified models are "orientation-based" models, meaning that the type of AGN that we see depends on the angle from which we are viewing it, as described above.

At least two types of unification models are used: radio-quiet and radio-loud. The radio-quiet unification models work with Seyfert galaxies. For Seyfert I's, the observer is looking directly at the active nucleus. For Seyfert II's the active nucleus is obscured by some structure, perhaps an opaque torus that prevents a direct view. There are two regions as defined by the type of emission lines they produce: the broad-line region that originates from cold material close to the black hole; and the narrow-line region, which comes from more distant cold material, which is why they are narrower. The Seyfert II's situation has the torus blocking the broad emission line region but leaving some of the narrow line region in view. At higher luminosities, quasars take the place of Seyfert I's (but there are no Quasar II's).

Radio-loud unification has historically involved high-luminosity radio-loud quasars, which can be unified with narrow-line radio galaxies, just like the Seyfert I's and II's are unified. As before, the viewing angle determines what type of radio-loud object we see. At very small angles to our line of site, relativistic beaming dominates the view, and we see a blazar. In contrast to this high-luminosity object, the population of radio loud galaxies is completely dominated by low-luminosity objects. These, unlike the high-luminosity objects, cannot be unified with quasars.

Seyfert Galaxies

Seyfert galaxies are small spiral galaxies that have small, very luminous nuclei. One of the differences between Seyferts, AGN's, and normal galaxies is the spectrum that each produces. Normal spectra include the combined light of many millions of stars, but the Seyfert spectra feature an overlay of broad emission lines on a normal spectrum. The emission lines are produced by highly ionized atoms (hydrogen, helium, nitrogen, and oxygen) being carried along at velocities between 500 and 4,000 km/s (with velocities up to 10,000 km/s) near the center of the galaxy (which can be as much as 30 times larger than the velocities near the center of normal galaxies). These high velocities result from the gravitational influence of a supermassive central black hole "weighing in" up to a few billion solar masses.

Seyfert galaxies also display strong emission in the infrared, ultraviolet, and X-ray part of the spectrum, and only 5% of Seyferts are radio-loud. The radio emission is thought to originate from synchrotron emission from the jet. The infrared emission arises from radiation in other bands that are being reprocessed by dust near the nucleus.

Some 2% of the spiral galaxies in the universe are Seyferts, which come in two varieties. Type 1 Seyferts are very luminous in the X-ray and ultraviolet parts of the electromagnetic spectrum. Their spectra show broad emission lines (fast-moving gas) with sharp, narrow cores (low velocity gas). Type 2 Seyferts lack the strong X-ray emission and have emission lines that are narrower than the Type 1's but broader than a normal galaxy. In both types the nuclei fluctuate in brightness rapidly, especially in X-ray "light" – this points to the sources being quite tiny, about the size of the local Solar System out to Neptune's orbit.

It is possible that Seyfert galaxies got to be the way they are through interactions or collisions or both. And then there are the BL Lac objects, which are a type of active galactic nucleus named after their prototypes BL Lac. These objects are characterized by their rapid variability (the prototype object was originally thought to be a variable star) and significant polarization in the optical range. This is in contrast with other AGN, which lack these features.

Quasars

Quasars (short for "Quasi-Stellar Objects" or QSO's) are faint, star-like points of light that have peculiar emission spectra [76]. These were discovered in the 1960s and are known as some of the most distant objects in the universe. The first quasar to be discovered was an object now known as 3C 48; the brightest quasar in the sky, 3C 273, was discovered shortly thereafter.

Quasars are the extremely bright nuclei of normal galaxies, as images from the Hubble Space Telescope (Fig. 5.2) would eventually show. The quasar 3C 273 appears to be part of a giant elliptical galaxy and is observed to eject a jet much like those in active galaxies. In some cases, quasars are located in galaxies that are parts of larger clusters of galaxies. Quasars are strong sources of radio emissions and are very far away; they are visible from such vast distances because they have the luminosities of between ten and one thousand times that of a large galaxy.

All observed quasar spectra display a redshift between 0.056 and 7.085 which translates to between 600 million and 28 billion light years away (184 million–8.6 billion parsecs, in terms of proper or fixed [not expanding] distance). Most quasars are farther than three billion light years and so they appear faint visually. The brightest quasar in Earth's sky is 3C 273 in Virgo, with an average apparent magnitude of 12.8. If it were able to be brought to 33 light years (10 pc) distance, it would have an apparent magnitude of −26.7, or about as bright as our Sun's apparent magnitude. This is defined as the object's absolute magnitude. The intrinsic luminosity of the quasar is about two trillion times that of the Sun, or about 100 times that of the total light of average giant galaxies such as the Milky Way.

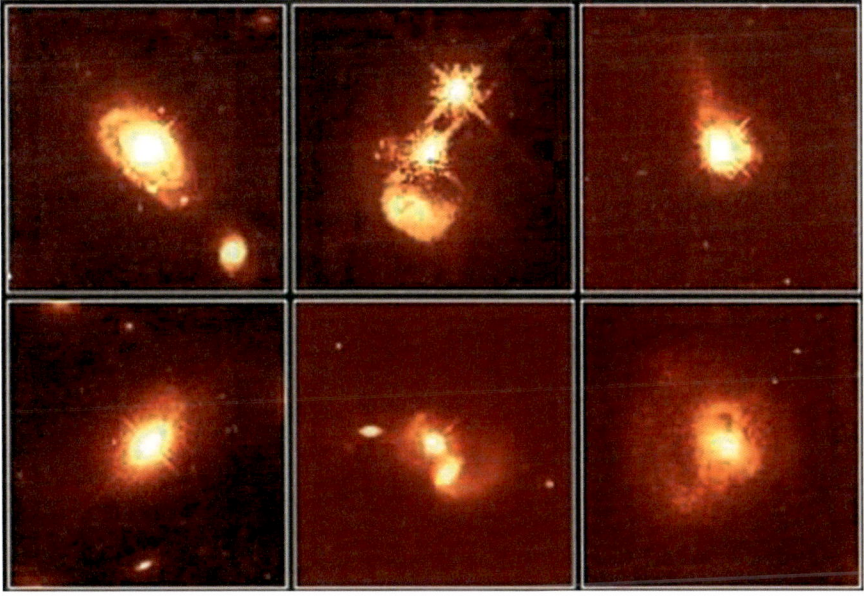

Fig. 5.2. Some active galaxies (Courtesy NASA and STScI) Object Names: PG 0052 + 251, PHL 909, IRAS04505-2958, PG 1012 + 008, 0316-346, IRAS13218 + 0552 (Image credits: John Bahcall [Institute for Advanced Study, Princeton], Mike Disney (University of Wales), and NASA, hubblesite.org [79]).

It is likely that the radiation is not radiating equally in all directions but is radiating in the form of two oppositely directed jets. If our line of sight is aligned with the axis of one of the jets, the object would appear much brighter than if we are looking at it from a vantage point perpendicular to the axis of the jets.

Another behemoth is the hyper-luminous quasar APM 08279 + 5255, which is brightened by gravitational lensing. The object has been magnified by a factor of 10 or so, but even in its un-enhanced form, it is still considerably more luminous than nearby quasars like 3C 273. The Ursa Major double quasar, 0957 + 561, was the first object determined to be distorted by a phenomenon later known as gravitational lensing. Astronomers were able to tell that this was two images of the same object (as opposed to a binary or binary-like object) because the spectral patterns were virtually identical. This object is one of the targets presented in Chap. 10.

Quasars typically have a luminosity of 10 [41] Watts or Joules per second. In order to generate this much energy, a supermassive black hole at the core of the quasar would need to consume the equivalent of at least ten Sun's worth of material per year. The brightest quasars known consume about 1,000 solar masses worth of material each year; the largest known quasar gorges itself on an estimated 600 Earths per minute worth of material. Depending on the availability of the material, quasars can "turn on and off"; once a quasar has consumed all the material within its reach, it turns off and becomes a normal galaxy.

There are more than 200,000 quasars known to date, most from the Sloan Digital Sky Survey. They show evidence of containing elements heavier than helium via their spectra. In order for this to be the case, the host galaxies must have experienced a massive rate of star formation, generating Population III (almost metal-free) stars sometime before the first observed quasars. Population III stars are yet to be confirmed, but some evidence for their existence may have been observed with NASA's Spitzer Space Telescope in 2005.

Quasars, like all unobscured active galaxies, can be strong X-ray sources; these can also be radio-loud objects that produce X-rays and gamma rays by a process known as inverse Compton scattering. A thermal spectrum is produced by the accretion disk surrounding a black hole, and the lower energy photons from this spectrum get scattered to higher energies by radio-emitting, relativistic electrons in the jet and surrounding corona. This leads to the production of the observed X-rays and gamma rays.

Other Exotic Objects

Blazars

A blazar (blazing quasi-stellar object) is a very compact quasar at the center of an active, giant elliptical galaxy [77]. These objects are among the most energetic known and are a type of active galactic nucleus. The ultimate power source of a blazar, like all AGN, is material falling onto a supermassive black hole at the center of a host galaxy. Blazars have relativistic jets pointing in the general direction of Earth, so that we are looking "down" the jet. This viewing perspective explains the rapid variability and compact features of two types of blazars, the optically violent variables (OVV) and BL Lac objects. Many blazars display superluminal features, that is, features that appear to be traveling at faster than light speed, but this is an illusion brought on by viewing perspective.

Some examples of blazars include Markarian 321, Markarian 501, 3C 454.3, 3C 273, BL Lacertae, and 2155-304. The two Markarian objects are also called "TeV blazars" due to their high energy gamma-ray emission in the tera-electron-volt range.

BL Lacertae

A BL Lac object is a type of active galaxy with an AGN and is characterized by rapid, large-scale flux variability and significant (a few percent) optical polarization [78]. The prototype (BL Lac) was thought to be an ordinary variable star due to its rapid variations. When compared with the more luminous AGN with strong emission lines, BL Lac objects have a featureless, non-thermal continuum.

The unified picture of AGN interprets BL Lacs as being due to a relativistic jet that is closely lined up to the line of sight of the observer. BL Lacs are likely identical to low-power radio galaxies. These AGN are hosted in massive spheroidal galaxies and are classified as blazars. All the known BL Lacs are associated with core-dominated radio sources, many of them in apparent superluminal motion. Some examples of BL Lac objects are BL Lacertae itself, OJ 287, AP Librae, PKS 2155-304, PKS 0521-365, Markarian 421, 3C 371, W Com, ON 325 and Markarian 501.

Optically Violent Variables

An optically violent variable quasar (OVV quasar) is a kind of highly variable quasar that also is a subtype of a blazar. The populations of objects that make up this class of object are few: these are rare, bright radio galaxies whose optical light output can change by as much as 50% in a single Earth day, hence the descriptive term. These objects are also similar to BL Lacertae objects in terms of appearance, but they are distinguished by stronger broad emission lines in their spectra and higher red shift components.

Why Are Faint Objects so Faint?

There are several reasons that can account for the faintness of a given object. If an object such as a galaxy is physically small or has low surface brightness, then the object will appear quite faint in Earth's skies. Many of the dwarf irregular and dwarf elliptical galaxies fit this description, as do faint areas of emission and reflection nebulae. The faintness can be further enhanced if interstellar material gets between us and the object. There are several examples of galaxies that would appear at a respectable brightness were it not for the location of the galaxy near the plane of our own Milky Way galaxy, where interstellar material and stars combine to nearly "erase" the galaxy from our view.

Star clusters, if sparse enough, begin to blend into the rich star fields of the Milky Way. These clusters, if located far enough from the plane of the galaxy, are

Fig. 5.3. NGC 6946 in Cepheus, an example of a galaxy dimmed by Milky Way material (Image courtesy of Paul and Liz Downing).

easier to see, but if they are sufficiently far enough away, they appear quite diffuse and faint. The same is true for globular clusters, which come in a range of sizes and concentrations. Such clusters may be so anemic as to appear as little more than a rich open cluster; place one of these "in front of" a rich star field and they become nearly lost in the celestial scenery. Several of the "Palomar" globulars that are featured in the globular cluster observing target list in Chap. 9 are like this, and M71 is a bit like this (but not nearly as challenging as the Palomars). Then globulars (even the nice ones similar to M13 and M22) may be faint due to great distances, which is the case for globular clusters belonging to nearby galaxies, and the most distant Milky Way globular clusters. These are also mentioned in the latter half of this book.

Nebulae are generally faint objects to begin with, with some notable exceptions, but some are much fainter than others. Planetary nebulae come in a stunning array of shapes, sizes, and colors, and provide an excellent example on the difference between surface brightness and apparent magnitude. There are objects such as the Helix Nebula that are large and have a relatively bright magnitude, but its low surface brightness renders it invisible in less than ideal sky conditions. Other nebulae may be so small that they blend in with the background star field. They themselves may have a relatively high surface brightness, but their small angular sizes (rendering it stellar except at the highest of magnifications) make them challenging objects to identify.

There are instances whereby an object can be intrinsically bright and large but far away or placed behind a veil of interstellar extinction, which results in the object appearing faint in our skies (Fig. 5.3). The vast majority of galaxies accessible to backyard telescopes are faint due to distance. Objects that are well within reach of CCD imagers and the Hubble Space Telescope provide a significant challenge for visual observers with large 'scopes. A sample of such objects in the format of 16 tables, is provided in Chap. 10.

Chapter 6

Ground- and Space-Based Observations of the Most Distant Parts of the Universe

Introduction

Let's take a look at some of the "starships" of our time, facilities that have given us the deepest views of the universe ever seen by humankind. The "starships" referred to here are the largest ground-based telescopes, the Hubble Space Telescope, and the other three members of NASA's "Great Observatories" (Spitzer Infrared Space Telescope, Chandra X-ray Observatory, Compton Gamma Ray Observatory), all of which have given us our deepest views of the universe yet. They are called "starships" because of their ability to "take" us to these distant places through extreme imaging, either looking out in wavelengths invisible to the human eye or collecting sparse amounts of photons over (for example) a 30-h period to reveal the most distant galaxies that would otherwise be invisible.

No doubt many will argue that other great observatories should be considered as well, but it is not necessary here to go too deeply into this subject. The "Four Great Observatories" are noteworthy in themselves because of their accomplishments and unique vantage points; the Keck telescopes are also selected because of their sheer size and the uniqueness of putting two large telescopes side-by-side, equipped with adaptive optics to provide an even sharper, more powerful view of the universe. Each of these five can "see farther" than any of the telescopes on Earth (or off of Earth). They also serve as representatives, at least for the purpose of this writing, of the other great observatories on Earth or in space, both in the present and in the near future.

In addition we will (very briefly) touch upon some of the astronomical surveys that are current as of this writing. Surveys and catalogs date back to the ancient Babylonians and continued with the likes of Messier and Herschel, but these persist to this day as we are able to peer deeper and deeper into the cosmos. One of the most interesting catalogs being assembled on an ongoing basis is that of the known and candidate exoplanets (also known as extrasolar planets) which is topping 3,000 if one includes all the candidates the Kepler satellite has amassed. Exoplanets are beyond the scope of this book, so we will focus on catalogs and surveys of non-planetary objects beyond the local Solar System.

B. Cudnik, *Faint Objects and How to Observe Them*, Astronomers' Observing Guides,
DOI 10.1007/978-1-4419-6757-2_6, © Springer Science+Business Media New York 2013

The Keck Observatory

The W. M. Keck Observatory is situated near the summit of a volcanic island perched at an elevation of 13,600 ft (4,145 m) on Mauna Kea, Hawaii [80]. The observatory features two telescopes side-by-side, each with primary mirrors of 33 ft (10 m) across. The Keck telescopes are the second largest optical telescopes in the world, second only to the Gran Telescopio Canarias (10.4 m or 410 in. across) at the Roque de los Muchachos Observatory on the island of La Palma among the Canary Islands of Spain.

The first telescope known as Keck I became operational in 1993, and the second telescope, Keck II, saw first light in 1996. Both telescopes are equipped with adaptive optics that correct for the blurring of the atmosphere because of turbulence.

The two telescopes can work together as the Keck Interferometer, which is capable of providing an effective aperture of 85 m (276 ft). This corresponds to a spatial resolution of 5 milli-arcseconds (mas) at 2.2 μm and 24 mas at 10 μm. This setup cannot make two-dimensional images because of the lack of a third telescope situated in a different direction, but it can make angular diameter measurements.

The telescopes are equipped with a suite of instruments that include both cameras and spectrometers, all of which work across much of the visible and near infrared parts of the electromagnetic spectrum. These instruments include:

- *DEIMOS* (*D*eep *E*xtragalactic *I*maging *M*ulti-*O*bject *S*pectrograph) can gather spectra from 130 (or more) galaxies in a single exposure. In "Mega Mask" mode, this number jumps to more than 1,200 using a special narrow-band filter.
- *HIRES* (*HI*gh *R*esolution *E*chelle *S*pectrograph) is the largest and mechanically most complex of the Keck's main instruments. This instrument breaks up starlight into its component colors and precisely measures the intensity of each of thousands of color channels. With a radial velocity precision up to one meter per second, this instrument has detected more exoplanets than any other in the world.
- *LRIS* (the *L*ow *R*esolution *I*maging *S*pectrograph) is a faint-light instrument that is able to take spectra and images of the most distant known objects in the universe. It can explore stellar populations of distant galaxies, clusters of galaxies, quasars, and AGN.
- *NIRC* (the *N*ear *I*nfra*R*ed *C*amera) is used on the Keck I telescope and is so sensitive it can detect a single candle flame on the Moon. It is ideal for ultra-deep studies of galactic formation and evolution; it can search for proto-galaxies and take images of quasar environments. It has also been used to study the galactic center, protoplanetary disks, and high-mass star-forming regions.
- *NIRC-2* (the second generation *N*ear *I*nfra*R*ed *C*amera) works with the adaptive optics system to produce the highest resolution images from any ground-based instrument operating in the 1–5 μm range. Research involves mapping surface features on Solar System bodies, searching for planets around other stars, and analyzing the morphology of distant galaxies.
- *NIRSPEC* (the *N*ear *I*nfra*R*ed *SPEC*trometer) observes very high redshift radio galaxies, the motions/types of stars near the center of the Milky Way, the physical nature of brown dwarfs, and the nuclear regions of dusty starburst

Fig. 6.1. The Keck II telescope showing its segmented mirror (Image courtesy of SiOwl [Andrew Cooper], who provides the image under the Creative Commons Attribution 3.0 Unported license).

galaxies, AGN, interstellar chemistry, the physics of stars, and research related to the Solar System.

- *OSIRIS* (the *O*H-*S*uppressing *I*nfra*R*ed *I*maging *S*pectrograph) is a spectrograph operating in the near-infrared and is used with the adaptive optics system. This instrument takes spectra in a small field of view in order to provide a series of images at different wavelengths. It enables suppression of emission of OH due to Earth's atmosphere, which increases the sensitivity to an extent that objects ten times fainter than previously available are detected.

The Keck Observatory is managed by the California Association for Research in Astronomy, whose board of directors includes individuals from the University of California and Caltech. Telescope time is allocated by the partner institutions (Caltech, the University of California, the University of Hawaii and its system of campuses) and is granted to their researchers. NASA and NOAO (National Optical Astronomy Observatory) accepts proposals from U. S.-based researchers and researchers around the world, respectively (The Keck II telescope is shown in Fig. 6.1).

The Hubble Space Telescope

Of all the observatories talked about here, no one has captured the public's collective heart more than the Hubble Space Telescope. This iconic "starship" of the twentieth and early twenty-first century has given the general public thousands upon thousands of beautiful images of unprecedented clarity, beauty, and scientific yield. From the Moon to the most distant galaxies ever seen, HST has, over the past 22 years, delivered over and above the call of duty. Although it was not the first telescope to be launched into space, it is one of the largest and most versatile telescopes to use space as its platform. It is one of four of NASA's "Great Observatories." The Compton Gamma Ray Observatory is no longer in operation, the Chandra X-ray Observatory is still operational, and the Spitzer Space Telescope is also still operational but no longer at 100%.

Hubble Space Telescope (HST) was carried into space in April 1990, aboard the space shuttle Discovery (STS 31) [81]. It was placed in low Earth orbit (but higher than the orbit of the International Space Station and the shuttles when they still flew) on April 24, 1990. The telescope contains a 2.4-m (7.9 ft) aperture mirror and four main observing instruments that operate in the near UV, the visible, and the near IR. Hubble was used to assemble the Ultra Deep Field (UDF) image, which is the most detailed, deepest visible light image ever made of the most distant parts of the universe. In fact, HST has been instrumental in many astrophysical breakthroughs, the most significant of which is the accurate determination of the rate of expansion of the universe.

The idea for the Hubble Space Telescope goes back to 1923, when three individuals, Hermann Oberth, Robert H. Goddard, and Konstantin Tsiolkovsky, published a paper that laid out in detail how a rocket could be used to loft a telescope into Earth orbit. In 1946, the astronomer Lyman Spitzer (after whom the telescope was named) wrote a paper about the "astronomical advantages of an extraterrestrial observatory." In fact, Spitzer spent most of his career pursuing the goal of having a space telescope built and put into orbit, and his persistence paid off with his appointment as head of a committee to define the scientific objectives for a large space telescope. The vision was one step closer to reality with the launch of the Orbiting Solar Observatory by NASA in 1962. The purpose of this first space-based solar observatory was to obtain spectra in the ultraviolet, X-ray, and gamma-ray parts of the spectrum. The United Kingdom followed up with its own orbiting solar telescope that same year.

The first orbiting observatory for "nighttime" purposes, the Orbiting Astronomical Observatory (OAO-1), was launched in 1966, but its battery failed after only 3 days, bringing an abrupt end to that mission. In 1968, OAO-2 was launched and worked just fine for nearly 4 years until 1972, which was much longer than the 1-year original planned lifetime of the mission. This space-based observatory made observations of stars and galaxies in the UV part of the spectrum.

The development of the mission that would become known as the Hubble Space Telescope continued in the 1970s and 1980s. Funding for this project was sought in the 1970s and was secured by 1978. At this time, the launch date for this large orbiting observatory was set for 1983. It was named after the astronomer Edwin Hubble, and like its human namesake and predecessor, the primary task was to study the expansion of the universe and to pin down an accurate estimate of H_0, the Hubble constant. It would do this by looking for and observing Cepheid variable stars in

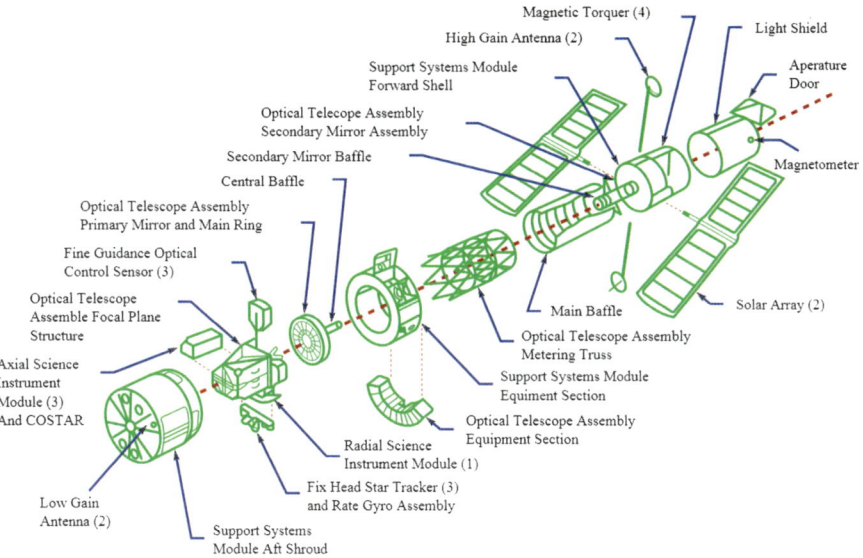

Magnetic Torquer (4)
Light Shield
High Gain Antenna (2)
Aperature Door
Support Systems Module Forward Shell
Optical Telecope Assembly Secondary Mirror Assembly
Secondary Mirror Baffle
Magnetometer
Central Baffle
Optical Telescope Assembly Primary Mirror and Main Ring
Fine Guidance Optical Control Sensor (3)
Optical Telescope Assemble Focal Plane Structure
Main Baffle
Optical Telescope Assembly Metering Truss
Axial Science Instrument Module (3) And COSTAR
Support Systems Module Equiment Section
Solar Array (2)
Optical Telescope Assembly Equipment Section
Radial Science Instrument Module (1)
Fix Head Star Tracker (3) and Rate Gyro Assembly
Low Gain Antenna (2)
Support Systems Module Aft Shroud

Fig. 6.2. Exploded view of HST with its components depicted (Hubble Space Telescope Servicing Mission 3A Media Reference guide, courtesy of NASA).

galaxies as far away as those in the Virgo Supercluster (about 65 million light years or 20 Megaparsecs). Work on the project was divided among many institutions, which included NASA's Marshall Space Flight Center in Alabama, NASA's Goddard Space Flight Center in Maryland, Perkin-Elmer, and Lockheed.

The optical telescope assembly (an exploded view of the components is shown in Fig. 6.2) was constructed similar to large professional telescopes (Cassegrain-type reflectors of Ritchy-Chretien design). The mirror was to be polished to 1/65 the wavelength of red light (the most accurate ever polished); its construction began in 1979 and was completed in 1981. The telescope was designed to be diffraction-limited in terms of resolution, 0.05 arc seconds for the HST 94-in. (2.4-m) primary in space. This is much better than the 0.5–1.0 arc seconds that is the typical limited seeing for ground-based observatories with telescopes of comparable size. The telescope was to observe in the UV, visible, and IR parts of the spectrum, especially the parts of the UV and IR that are absorbed by the atmosphere.

Originally set to launch in 1983, the HST was delayed until October 1986 after several postponed launches. The *Challenger* space shuttle explosion of January 28, 1986, further delayed the launch of HST. Two and a half years later, the shuttle fleet returned to service, with the September 29, 1988, launch of the space shuttle *Discovery*, and in April 1990, STS-31 carried the HST into orbit.

Now that the space telescope was in orbit, testing could begin on the system, and science observations were soon to follow. However, the early returned images showed a serious problem with the optics system. The primary mirror had been ground to the wrong shape; although the mirror was still one of the most precisely figured mirrors ever made, it was too flat at the edge by about 2.2 μm. This almost imperceptible flaw introduced severe aberrations as the light reflecting from the edges of the mirror focused off center. The good news was that this was easily corrected. In December 1993, the first servicing mission of the Hubble Space Telescope was carried out, and it was very successful. The telescope was fitted with

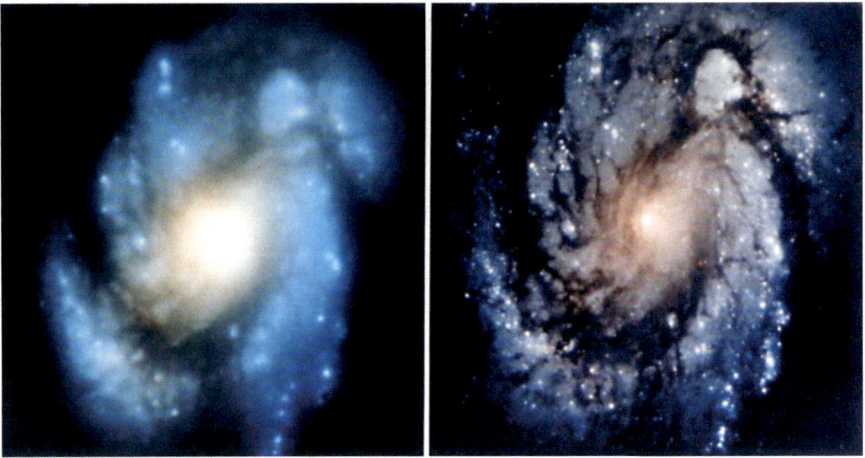

Fig. 6.3. Before and after images showing the degradation of the flawed optics and the corrected optics (Image courtesy of NASA).

a correcting lens that removed the aberrations. The difference in image quality before and after is nicely illustrated in Fig. 6.3.

The improvement of Hubble's imaging system was done in the first of five servicing missions; the last servicing mission was in 2009. During these flights, the solar arrays were replaced, equipment and gyroscopes were changed in and out, and other work was performed. After the second shuttle disaster of February 1, 2003, with the *Columbia* disintegrating on re-entry, killing all seven astronauts onboard, there was concern about sending astronauts to the elevated orbit of the HST, taking the crew and spacecraft far from the International Space Station. If another incident like the one that was found to have caused *Columbia's* demise (pieces of the external fuel tank casing broke off and hit the shuttle tiles, damaging them) were to occur on a servicing mission, the astronauts would not be able to easily "abandon ship," which was the recommended action of the day. Robotic missions were proposed that would do the servicing work, but ultimately humans would be given the go ahead to return to the HST with the May 2009 servicing mission, after a gap of over 7 years from the previous servicing mission, and the last one scheduled to be performed.

The list that follows highlights some of the major accomplishments of the Hubble Space Telescope. In addition to returning hundreds of thousands of spectacular images showing the universe in a way that humans had never seen before, a lot of quality science was performed by the HST and its teams of scientists. Here are some of these key accomplishments [82]:

- Narrowed down the age of the universe to between 13 and 14 billion years
- Key role in discovery of dark energy
- Shown galaxies at all stages of evolution
- More than 6,000 scientific articles published based on Hubble data
- Established the presence of black holes in the nuclei of nearby galaxies
- Discovered protoplanetary disks in the Orion Nebula
- Provided evidence for the presence of extrasolar planets around sun-like stars

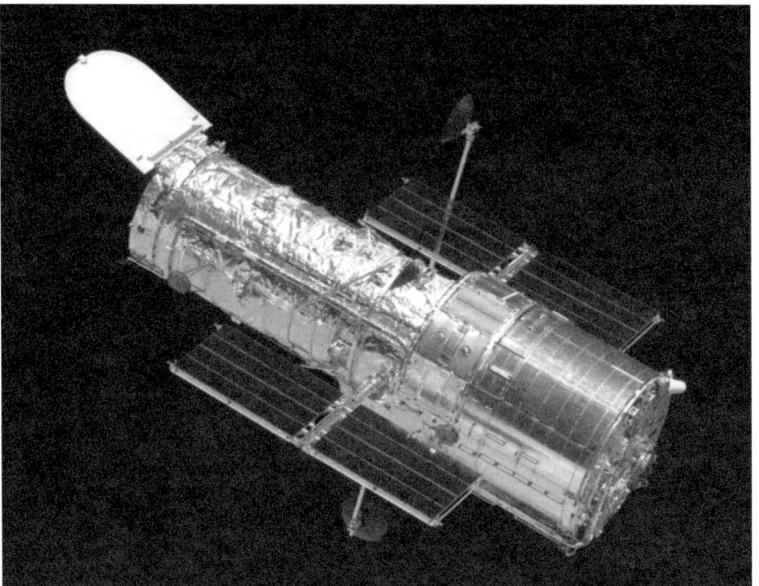

Fig. 6.4. The Hubble Space Telescope as it appeared during Servicing Mission 4 (Image courtesy of NASA and taken by Ruffnax [Crew of STS-125]).

- Found optical counterparts of still-mysterious gamma ray bursts
- Made a detailed study of Pluto and Eris
- Watched the impacts of Shoemaker-Levy 9 as it rammed into Jupiter

The Hubble Space Telescope (Fig. 6.4) is expected to continue operations until at least 2014 and perhaps beyond. Its successor, the James Webb Space Telescope, is expected to take its place, but as of late there were still some funding problems that threaten to seriously delay or even put the project on hold indefinitely. The JWST is a large infrared space telescope with a 6.5-m primary mirror. Currently, the launch of this space telescope is scheduled for 2018. More about how this instrument could change our view of faint objects will be presented later in this chapter.

The Chandra X-Ray Observatory

Formerly known as AXAF (the Advanced X-ray Astrophysics Facility), Chandra (Fig. 6.5) was launched on July 23, 1999, aboard shuttle mission STS-93 and has served as NASA's flagship X-ray observatory since then [83]. In order to study X-rays it was necessary for any X-ray observatory to be lofted above Earth's atmosphere, which is opaque to X-rays, hence the placement of Chandra into orbit. The orbiting observatory was renamed Chandra in 1998 in honor of Indian physicist Subrahmanyan Chandrasekhar, who is famous for determining the mass limit of white dwarf stars.

This mission was first proposed in 1976 to NASA by Riccardo Giacconi and Harvey Tananbaum. The work began at NASA's Marshall Space Flight Center

Fig. 6.5. Chandra inside the payload bay of space shuttle *Columbia* (Image courtesy of NASA and the Kennedy Space Center).

(MSFC) and the Smithsonian Astrophysical Observatory (SAO). The first imaging X-ray telescope was launched in 1978 and was named Einstein (HEAO-2), and this was collecting data from orbit as work continued on the space telescope formerly known as AXAF. The spacecraft was redesigned several times before the final assembly and testing took place with TRW (now Northrop Grumman Aerospace Systems) in Redondo Beach, California. Chandra was launched in 1999, carrying the distinction of being the heaviest payload ever brought up by the space shuttle *Columbia*, at 22,753 kg.

The instruments that are carried onboard the Chandra X-ray observatory include the following:

- The *Science Instrument Module* (*SIM*) includes two focal plane instruments, the *Advanced CCD Imaging Spectrometer* (*ACIS*), and the *High Resolution Camera* (*HRC*). The ACIS provides images and spectral information, and works between the energies of 0.2 and 10 keV (X-rays and gamma rays are typically referred to in terms of energies). The HRC provides images in the energy range 0.1–10 keV and has a cadence (time resolution) of 16 ms.
- The *High Energy Transmission Grating Spectrometer* (*HETGS*) works with photons that have the range in energy from 0.4 to 10 keV. The *Low Energy Transmission Grating Spectrometer* (*LETGS*) works from 0.09 to 3 keV.

Since its 1999 launch, the Chandra observatory has produced some significant findings about the X-ray universe. These findings have greatly advanced the field of X-ray astronomy and include the following:

- Scientists were able to get their first look at the compact object (perhaps a neutron star or a black hole) at the center of the supernova remnant Cassiopeia A.
- Chandra revealed a never before seen ring surrounding the Crab Nebula pulsar and jets that had, up to this point, been only partially revealed by earlier observations.
- Chandra "saw" the first X-ray emission ever observed from the supermassive black hole at the center of the Milky Way.
- For the first time, pressure fronts in detail were observed in the object named Abell 2142, which consists of merging clusters of galaxies.
- A new type of black hole was discovered in the active galaxy M82. The black hole is a sort of "missing link," containing a mass intermediate between stellar mass black holes and supermassive black holes.
- Chandra and BeppoSAX (an Italian-Dutch X-ray astronomy satellite) observations have together indicated that the mysterious gamma ray bursts may occur in star-forming regions.
- Unexpectedly, it was discovered that nearly all stars on the main sequence emit X-rays.

Fig. 6.6. A young pulsar only 12 miles across is responsible for this 150 light-year-long nebula seen in X-rays. The lowest energy X-rays are shown as *red,* the medium range X-rays are depicted as *green,* and the highest energy X-rays are shown in *blue.* The pulsar, PSR B1509-58, is thought to be 1,700 years old and situated about 17,000 light years distant (Image courtesy of NASA/CXC/CfA/P and Slane et al.).

- The Hubble constant was determined to be 76.9 km/s/Mpc using the Sunyaev-Zel'dovich (SZ) effect. (The SZ effect results from high-energy electrons making distortions in the cosmic microwave background radiation through inverse Compton scattering, like what happens in quasars).
- Chandra imaged the enigmatic region known as PSR B1509-58; this image, due to its appearance, became known as the famous "Hand of God" object (Fig. 6.6).

The Spitzer Space Telescope

This significant infrared space observatory (the fourth and final of NASA's "Great Observatories") was launched in 2003 [84]. Formerly known as the Space Infrared Telescope Facility (SIRTF), the facility was renamed after successful demonstration of operation (Fig. 6.7). It was originally expected to have a 2.5-year mission life, with an extension to 5 or more years until the liquid helium needed to cool the instruments onboard ran out. Without the liquid helium the instruments become useless, as they are swamped with the infrared emissions of their immediate surroundings and cannot see into the depths of the universe.

The liquid helium did run out on May 15, 2009, but the two shortest wavelength modules of IRAC (see below) are still able to operate with the same sensitivity as before and will continue to be used during the Spitzer Warm Mission (the extension of the Spitzer mission).

These are the three instruments that Spitzer carries on board:

- *IRAC* (*I*nfra*R*ed *A*rray *C*amera), an infrared camera that operates simultaneously on four wavelengths (3.6, 4.5, 5.8, and 8 μm), and each uses a 256×256 pixel detector. The first two are still fully functional at the telescope equilibrium temperature of about 30 K.
- *IRS* (*I*nfra*R*ed *S*pectrograph) is an infrared spectrometer with four submodules, each of which uses a 128×128 pixel detector. They operate at the following wavelengths: 5.3–14 μm (low resolution), 10–19.5 μm (high resolution), 14–40 μm (low resolution), and 19–37 μm (high resolution).
- *MIPS* (*M*ultiband *I*maging *P*hotometer for *S*pitzer) consists of three detector arrays in the far infrared (128×128 pixels at 24 μm, 32×32 pixels at 70 μm, 2×20 pixels at 160 μm). The 24 μm detector is the same one used as in the IRS short wavelength modules.

Some of the first images (whose purpose was to demonstrate Spitzer's capabilities) included galaxies, stellar nurseries, dusty disks (likely in the process of forming planets), and organic material in the distant universe. In 2005, the Spitzer Space Telescope was the first instrument to directly image light from extrasolar planets HD 209458b and TrES-1, two "hot Jupiters." Spitzer also found, in April 2005, that an object known as Cohen-kuhi Tau/4 had a planetary disk much younger and containing less mass than previously theorized. In addition to all this, Spitzer found a faintly glowing object in 2004 that may be the youngest star ever observed. It consists of a core of gas and dust referred to as L1014, with

Fig. 6.7. Spitzer telescope (Image courtesy of NASA-JPL-CALTECH).

a young star collecting and heating up gas and dust from the cloud surrounding it (Four examples of images from the Spitzer telescope are shown in Fig. 6.8).

The telescope is also being used to support what is called the Gould Belt Survey, which is observing Gould's Belt in multiple wavelengths. Gould's Belt is a partial ring of stars in the Milky Way Galaxy that spans 3,000 light years (900 pc) and is tilted toward the galactic plane by 16–20°. This may represent the spiral arm where the Sun resides (the Sun is currently situated 325 light years or 100 pc from the arm's center). The telescope is also participating in the GLIMPSE and MIPSGAL surveys. GLIMPSE, the Galactic Legacy Infrared Mid-Plane Survey Extraordinaire, is a survey of 300° of the inner Milky Way Galaxy, made up of approximately 444,000 images in four separate wavelengths using IRAC. MIPSGAL is similar, covering 278° of the Milky Way's disk at longer wavelengths. Astronomers revealed the largest, most detailed infrared picture of the Milky Way yet created (with more than 800,000 individual pictures) on June 3, 2008, at the 212th meeting of the American Astronomical Society in St. Louis, Missouri.

Fig. 6.8. Examples of Spitzer's handiwork: going clockwise from top left: IR view of the galaxy M81, embedded outflows from the protostar Herbig-Haro 46/47, a dark globule in IC 1396 showing protostars, and Comet Schwassmann-Wachmann 1 (Images courtesy of NASA).

The Compton Gamma Ray Observatory

The Compton Gamma Ray Observatory (CGRO, Fig. 6.9) was the second of the four "Great Observatories" operating from Earth orbit (at 280 miles or 450 km altitude) between 1991 and 2000 [85]. The orbiting observatory detected light with energies ranging from 20 keV to 30 GeV (0.02–30,000 MeV), covering the x-ray and gamma ray portion of the electromagnetic spectrum. The instrument was launched via the space shuttle *Atlantis* on April 5, 1991, and de-orbited on June 4, 2000.

The observatory was named in honor of Dr. Arthur Holly Compton of Washington University in St. Louis, who was a Nobel prize winner for his work with gamma ray physics. CGRO was built by TRW (which is now Northrop Grumman Aerospace Systems) in Redondo Beach, California. The project was an international collaboration, involving, among many others, the European Space Agency, various universities, and the U.S. Naval Research Observatory.

Fig. 6.9. Compton Gamma Ray Observatory (Image courtesy of NASA).

The four instruments carried on board CGRO covered an unprecedented six orders of magnitude of the electromagnetic spectrum. The instruments are described in order of increasing spectral energy coverage:

- *BATSE*, the *B*urst and *T*ransit *S*ource *E*xperiment searched the sky for gamma ray bursts (20–600 keV) and carried out full sky surveys to look for long-lived transient sources. It is made up of eight identical detector modules, one at each of the satellite's eight corners. Each module includes a Large Area Detector (LAD) that covers the 20 keV to ~2 MeV range, and a Spectroscopy Detector, extending the upper energy range to 8 MeV. These are the instruments that observe the gamma ray bursts that have been detected at roughly one per day over the 9-year mission lifetime.

- *OSSE*, the *O*riented *S*cintillation *S*pectrometer *E*xperiment, provided by the Naval Research Laboratory, detects gamma rays that enter the field of view of any of four detector modules. Each module can be pointed independently at a desired region and are effective in the 0.05–10 MeV range. These detectors can work in pairs during a gamma ray event with one measuring the event itself and the second measuring the background, then switching roles to provide more accurate measurements of both source and background.

- *COMPTEL*, the Imaging *COMP*ton *TEL*escope, by the Max Planck Institute for Extraterrestrial Physics, the University of New Hampshire, the Astrophysical Division of the European Space Agency, and the Netherlands Institute for Space Research, works in the 0.75–30 MeV. This instrument is able to determine the direction of a gamma ray burst within a degree and the energy of the event to within a 5% accuracy (at the higher range of energies).

- *EGRET*, the *Energetic Gamma Ray Experiment Telescope* measures high energy gamma ray source positions to an accuracy of a fraction of a degree and photon energies with an accuracy of 15%. The energy range this operates in is from 0.02 to 30 GeV and was developed by Stanford University, Goddard, and the Max Planck Institute for Extraterrestrial Physics.

The instruments worked to complete various surveys of the entire sky, including the high-energy whole sky survey (with the EGRET instrument, mapping energy sources above 100 MeV) that, over four years, discovered 271 sources (170 of these were unknown prior to the survey). The COMPTEL instrument did an all sky map of aluminum-26 (a radioactive isotope of aluminum); the BATSE was the instrument that made the news on many occasions, making approximately 2,700 detections including at least one (GRB 080319B) that had a naked-eye (peak apparent magnitude of 5.8) optical counterpart. The frequency was one per day, and the vast majority originated outside of the Milky Way in distant galaxies, showing these events come from energetic sources. CGRO also completed an all-sky survey of pulsars and supernova remnants and discovered a terrestrial source of gamma rays: thunderclouds.

Our knowledge of gamma ray bursts was greatly expanded by CGRO. We are now confident that GRBs come in two forms: the short duration kind that last for 2 s, and the long duration kind that last longer than two seconds. Also, the first four examples of soft gamma ray repeaters were discovered. These sources are relatively weak (energies below 100 keV) and feature unpredictable periods of activity and inactivity. GRB 990123 was an important find, being one of the brightest bursts recorded and the first to include an optical afterglow that was observed, allowing astronomers to measure a redshift of 1.6 (distance of 3.2 billion parsecs or 10 billion light years). This event showed that GRB afterglows resulted from highly collimated beams directed Earthward, and not uniformly in all directions, which made the needed energy budget for such events more plausible.

The observatory was de-orbited for safety reasons after one of its gyroscopes failed. If a second were to fail it would have made de-orbiting much more difficult and the likelihood of a fall over populated regions more likely, so the decision was made to decommission the spacecraft right away. It made reentry on June 4, 2000, and the parts that survived reentry fell into the Pacific Ocean.

The Gran Telescopio Canarias

The Gran Telescopio Canarias (or "Canaries Great Telescope", Fig. 6.10) is currently the world's largest single-mirror reflecting telescope, with an aperture of 10.4 m (410 in.) [86]. It is part of the Roque de los Muchachos Observatory on the island of La Palma in the Canary Islands of Spain, and it saw first light on July 13, 2007. The institutes that collaborate on this project include the Instituto de Astrofisica de Canarias, the University of Florida, and the National Autonomous University of Mexico. The telescope operates at an elevation of 7,438 ft (2,267 m). The observing time is granted as follows: 90% Spain, 5% Mexico, and 5% for the University of Florida.

Preliminary observations happened as early as 2007 but the 'scope did not achieve the title of being the world's largest until 2009, since, at first light, only 12 segments

Fig. 6.10. The Gran Telescopio at Roque de los Muchachos Observatory, La Palma, Canary Islands (Image courtesy of Pachango, from the Wikimedia Commons).

of its primary mirror was used. There is now a total of 36 hexagonal segments that make up the primary mirror. The primary instrument is a detector by the name of OSIRIS, which is actually made up of three separate astronomical instruments. OSIRIS stands for Optical System for Imaging and low Resolution Integrated Spectroscopy, and it operates from 365 to 1,000 nm. It performs imaging and low-resolution spectroscopy (with long slit and multiobject spectroscopic modes).

The University of Florida provided an instrument known as the CanariCam, which is an imager that operates in the mid-infrared range and has spectroscopic, coronagraphic, and polarimetric capabilities. It is designed to be a diffraction-limited imager and provide a powerful and versatile addition to the instruments used on the telescope. The instrument will work in the thermal infrared part of the spectrum (between 7.5 and 25 μm) and was scheduled for installation in 2010.

Recent and Ongoing Astronomical Surveys and Future Plans

After having provided a brief history of astronomical surveys back in Chaps. 1 and 2 (the latter in conjunction with the biographical sketches of several of the astronomers that have been instrumental in expanding our view and our knowledge of the universe), we will now take a moment and bring the reader up to date on some of the current surveys that are going on as these pages are written. The universe is an unimaginably vast place, where large tracts of space remain unexplored. Current and future ground- and space-based observing initiatives are working to bring more and more of this uncharted cosmos into focus, not just with the optical part of the spectrum but also with other parts of the spectrum.

Table 6.1 summarizes the current, recent, and near future surveys that are happening on the cutting edge of astronomy. This is by no means a complete listing but gives a sizable sample of the activity, started by Messier and Herschel in the early history of the telescope and continuing to this day. The surveys highlighted work in the visible or infrared part of the spectrum.

Surveys have come a long way since the days of Messier and Herschel and their pioneering work. (In fact, there is an infrared space telescope named after Herschel that is operated by the European Space Agency; it is not included here since it does its work primarily in the far IR and microwave.) Even with this narrower focus, there are many projects that are ongoing to survey the heavens to varying degrees and through various wavelengths of the electromagnetic spectrum.

Ever since the lives of Messier and Hershel, we have discovered that the optical part of the electromagnetic spectrum was not all there is. The electromagnetic spectrum consists of a very broad expanse of radiation types, from the shortest wavelength gamma rays to the longest wavelength radio waves and everything in between. Surveys are using various parts of this spectrum to learn more and more about the vast universe we find ourselves in, and we are certainly finding amazing things.

We are not only using different forms of light to conduct our surveys, but we are also using unusual sources of power and exotic elements to aid our efforts to compile a census of what is out there. There are also dark-matter surveys that are underway, attempting to use every gravitational lensing occurrence, coupled with the best theories, to map out regions where dark matter is found. Projects such as CLASH (Cluster Lensing And Supernova survey with Hubble), which use the Hubble's keen vision to unlock the deepest, darkest secrets of the universe are opening up a new dimension in cosmic exploration.

Many additional survey projects are outlined at http://www.cv.nrao.edu/fits/www/yp_survey.html and show that much is going on these days in the area of faint object astronomy. Some of the interesting surveys on this list include several galaxy redshift surveys, one of which focuses on quasars. Many of these surveys are available to the public, and one of the citizen science projects (more on this in Chap. 12) involves users who will download and scrutinize images of galaxies and classify the galaxies that, prior to these surveys, were never before seen. The Hubble Deep Field survey is the deepest survey yet attempted, with the faintest of objects imaged for the first time ever.

The Giant Magellan Telescope

The next greatest ground-based telescope is on the drawing board as of this writing, and it is a huge one. The Giant Magellan Telescope (GMT) project is the work of an international consortium of top universities and research institutions. The GMT is expected to see first light by 2020 and will be located in Chile. The unique characteristic of the GMT will be its segmented mirror system. It will include not one but seven 8.4-m mirror that will form a single large mirror 24.5 m in diameter. This will result in the resolving power of the telescope being ten times greater than the resolving power of the Hubble Space Telescope.

Table 6.1. Select surveys that are recent or ongoing as of early 2012 [87]

Name of survey	Object(s) of focus	Facility(ies)	Lead astronomer(s) and institution(s) involved/comments
Pan Andromeda Archaeological Survey (PAndAS) 2008 to present (optical)	Andromeda Galaxy, M31 and the Triangulum Galaxy, M33	Canada-France-Hawaii Telescope	Dr. Alan McConnachie at the Herzberg Institute of Astrophysics (NRC-HIA), and involves over 25 investigators from that institute, as well as from universities in Canada, France, the United States, the United Kingdom, Germany, and Australia
Sloan Digital Sky Survey (SDSS) 2000–2014 (optical, near IR)	Galaxies, Quasars, Type Ia Supernovae, multi-filter imaging and spectrographic survey	2.5-m wide-angle optical telescope at Apache Point Observatory in New Mexico, United States	This is a large collaboration of about 150 scientists at a number of institutions. More information about the collaborators as well as the mission and the science yield is found at http://www.sdss.org/
The 2-μm All-Sky Survey (2MASS) 1997–2001, IR	All sky survey at the J, H, and K band (1.25, 1.65, and 2.17 μm), ground-based	Mt. Hopkins Arizona (northern half) and Cerro Tololo/CTIO Chile (southern half)	University of Massachusetts, Infrared Processing and Analysis Center (JPL/Caltech), NASA, and NSF
Akari (Astro-F) 2006–2008	A Japanese mid and far infrared all sky survey satellite	Akari, primary telescope aperture 27.0 in. or 68.5 cm	Japan Aerospace Exploration Agency, in coop with institutes of Europe and Korea
Wide-field Infrared Survey Explorer, December 2009 to July 2010	The purpose is to survey 99% of the sky at wave-lengths of 3.3, 4.7, 12, and 23 μm	WISE primary telescope aperture is 0.4 m (15.6 in.)	NASA/JPL with major contractors Ball Aerospace, Lockheed Martin, Space Dynamics Laboratory, and SSG Precision Optronics, Inc. Telescope more than 1,000 times as sensitive as previous infrared surveys. The initial survey, consisting of each sky position imaged at least eight times, was completed by July 2010
Galaxy and Mass Assembly (GAMA)	Large scale cosmic structure on scales of 1 kpc to 1 MPc, to include galaxy clusters, groups, mergers, and individual galaxies	Six listed in the next column	1. The Anglo-Australian Telescope (AAT) 2. The VLT Survey Telescope (VST) 3. The Visible and Infrared Survey Telescope for Astronomy (VISTA) 4. The Australian Square Kilometre Array Pathfinder (ASKAP) 5. The Herschel Space Observatory 6. The Galaxy Evolution Explorer (GALEX) More information at http://gama-survey.org/

(continued)

Ground- and Space-Based Observations

Ground- and Space-Based Observations

Table 6.1. (continued)

Name of survey	Object(s) of focus	Facility(ies)	Lead astronomer(s) and institution(s) involved/comments
Great Observatories Origins Deep Survey	This survey combines deep observations of mostly distant galaxies from three of NASA's "Great Observatories" (listed to the right) along with other space- and ground-based observatories	The facilities listed in the next column	GOODS data sources — space-based observatories: 1. The Hubble Space Telescope (optical imaging with the advanced camera for surveys) 2. The Spitzer Space Telescope (infrared imaging) 3. The Chandra X-Ray Observatory (X-ray) 4. XMM-Newton (an X-ray telescope belonging to the European Space Agency) 5. The Herschel Space Observatory (an infrared telescope belonging to the ESA) GOODS data sources — ground based observatories: 1. The Very Large Telescope (four 8.2-m telescopes) 2. Kitt Peak National Observatory 4-m scope

The telescope will operate in the wavelength range of 320–2,500 nm and have an effective resolution of 0.21–0.3" at 500 nm. The site has been selected, but construction has not begun on the observatory itself, although the mirrors are being built as of this writing. The configuration of the telescope will be one of Gregorian style, where there is a small circular hole in the primary mirror to allow light that has been reflected from the primary, back up the tube to the secondary mirrors near the front opening of the 'scope, then reflected back down the tube to the detector situated at the back. The advantage to this configuration is that a longer focal length can be "folded" into a more compact configuration.

The primary mirrors will be made at the Steward Observatory Mirror Laboratory in Tucson, Arizona, and will have a honeycomb structure to save on weight. The mirrors themselves will be precisely ground to within a few wavelengths of visible light, or about one-millionth of an inch. With the seven primary mirrors will be seven secondary mirrors, and these will be equipped with adaptive optics. Each of the flexible secondary mirrors will be equipped with hundreds of actuators that will continuously adjust the mirrors to cancel the effects of atmospheric turbulence. As a result, twinkling stars that appear as small fuzzy disks with magnification will be transformed into sharp, steady, pinpoints of light.

The location of the GMT will be advantageous as well, being situated in one of the highest and driest locations on Earth. This is within Chile's Atacama Desert, where skies are typically clear more than 300 nights per year. The elevation of the observatory will be over 8,500 ft (2,550 m) on a peak known as Las Campanas.

With all of this in its favor, the GMT will be well equipped to study faint objects, which will likely enable some of the universe's first galaxies to be seen for the first time. With a huge light-gathering capacity, coupled with advanced electronics and detectors, the GMT will go a long way in helping to answer some of the most fascinating questions in twenty-first century astronomy, such as: How did the first galaxies form and what do they look like? How did the universe make the first stars? What exactly is dark matter and dark energy, and how do these come into play to determine the structure and evolution of the universe? Are there other Earth's out there that are likely to support life? What is the fate of the universe?

Much more information about the telescope, the status of the project, who the consortium members are, and more can be obtained from the website www.gmto.org.

The James Web Space Telescope: the Ultimate in Faint Object Astronomy

Formerly known as the Next Generation Space Telescope (NGST), the project was renamed the James Webb Space Telescope (JWST, Fig. 6.11) in September 2002 after NASA's second administrator (who served this role from 1961 to 1968) [88, 89]. James Webb also played a role in the Apollo program and made scientific research a core activity of the NASA program.

The JWST will be a large space-based telescope that will operate in the near infrared part of the spectrum. As of this writing, the launch date is scheduled for sometime in 2018. The purpose of this space telescope will be to study the history of the universe, from the very first luminous glows after the Big Bang event to the formation of solar systems and the evolution of our own planetary system.

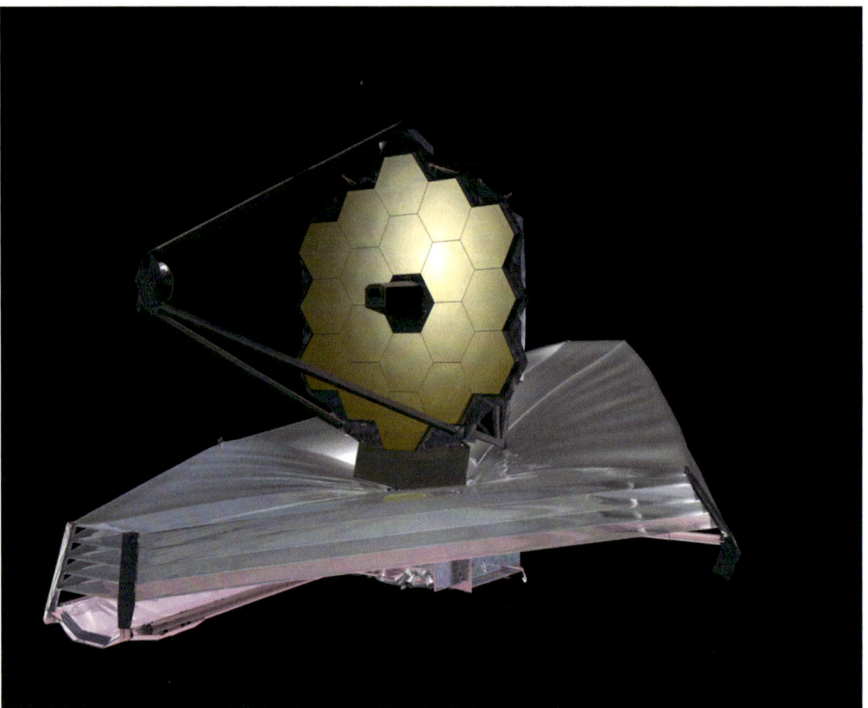

Fig. 6.11. The James Webb Space Telescope, as of September 2009 (Image courtesy of NASA).

The project originates from planning sessions dating back to 1993 for the successor of the Hubble Space Telescope. In fact, given the infrared nature of the observatory, it can be considered more a successor to Spitzer. The JWST will do far more than both combined, since it will be able to detect far more and much older stars and galaxies. Europe came on board with this project in 2007 with a ESA-NASA Memorandum of Understanding, which includes the launch vehicle (Ariane-5 ECA), several instruments (NIRSpec and MIRI Optical Bench Assembly), and manpower support. The telescope is planned to orbit the Sun about 930,000 miles (1,500,000 km) behind Earth at the L2 Lagrangian point (Fig. 6.13) moving at the same pace as Earth. It will actually be orbiting the L2 point in a halo orbit with a radius of 500,000 miles (800,000 km). This positioning will allow the JWST to use its radiation shield to protect it from the Sun's heat and light.

JWST will have about half the mass as the HST, but the former's mirror will be much larger, with a light collecting surface area some five times that of the HST (Fig. 6.12). The 'scope is designed to observe in the infrared for three main reasons: high-redshift objects have their visible light shifted to the infrared, cold objects (planets, debris disks) emit mostly in the infrared, and parts of the infrared cannot be readily studied by ground-based telescopes or existing space-borne instrumentation.

The science to be done by the JWST will include four main themes

· The End of the Dark Ages: First Light and Re-ionization
· The Assembly of Galaxies

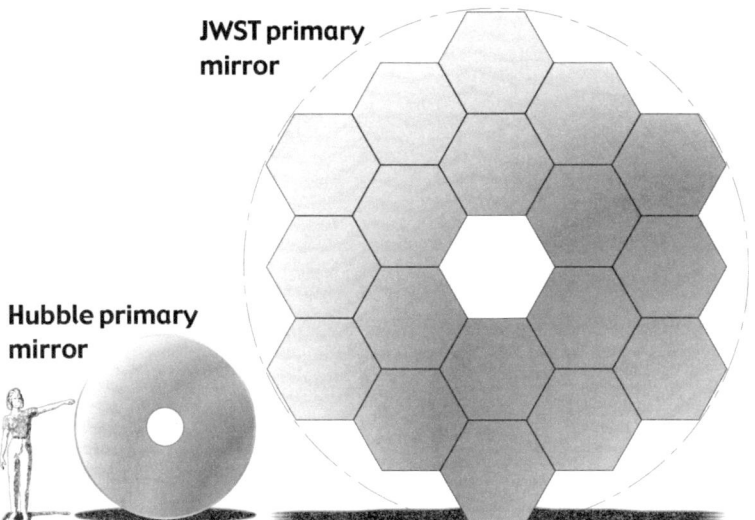

JWST primary mirror

Hubble primary mirror

Fig. 6.12. Size comparison between a human adult female, the Hubble primary mirror and the JWST composite mirror (Image courtesy of NASA).

- The Birth of Stars and Protoplanetary Systems, and
- Planetary Systems and the Origins of Life.

To achieve these objectives, the space observatory will be able to penetrate dusty objects, to reveal the activity within, in objects such as molecular clouds, the cores of active galaxies, and other dusty regions of space. The telescope will be able to image clouds in the local galaxy's interstellar medium, brown dwarfs in the solar neighborhood, and planets and Kuiper Belt objects in our own Solar System. On the other end of the scale, JWST's infrared capabilities are expected to reveal the very first galaxies in the history of the universe, forming a mere few hundred million years after the Big Bang.

A number of innovative technologies have been developed for JWST, including its folding segmented primary mirror, ultra-lightweight gold-coated beryllium optics, detectors sensitive enough to record extremely faint objects, microshutters enabling programmable object selection for the spectrograph, and a cooling system that brings the temperature of the mid-IR detectors down to 7 K. All of these technologies were successfully demonstrated by January 2007.

The telescope carries four detectors:

- *NIRCam* (*Near InfraRed Cam*era) is an imager that operates in the infrared, from 0.6 μm (in the orange portion of the visible) to 5 μm (in the near infrared); this instrument will also serve as the telescope's wavefront sensor, required to keep the mirror segments aligned.
- *NIRSpec* (*Near InfraRed Spec*trograph) will cover the same wavelength range as the NIRCam but will be doing spectroscopy in this range. The instrument will work in three observing modes: low resolution mode (using a prism), an R ~ 1,000 multi-object mode, and an R ~ 2,700 integral field unit or long-slit spectroscopy mode. The system will be capable of observing hundreds of individual objects in the field of view.

- *MIRI* (*Mid InfraRed Instrument*) covers the middle infrared wavelength range (5–27 μm), with a mid-IR camera and an imaging spectrometer.
- *FGS* (*Fine Guidance Sensor*) will be used to stabilize the alignment of the telescope during an observation. Both the orientation of the spacecraft and the mirror segment control for image stabilization will be controlled by the FGS. In addition, a Tunable Filter Imager (TFI) will be provided for astronomical narrow-band imaging in the 1.5–5 μm range. The TFI is part of the Near Infrared Imager and Slitless Spectrograph (FGS-NIRISS), all of which will be mounted with the FGS to make a single unit, but each instrument will serve a different purpose.

Each of the above instruments, except the FGS, is equipped with coronagraphs to block starlight to allow for observations of faint objects such as circumstellar disks and extrasolar planets, both of which occur next to bright stars.

The JWST is an international partnership between a number of countries and NASA, the European Space Agency (ESA), and the Canadian Space Agency (CSA). NASA Goddard manages the development effort; the prime contractor is Northrup Grumman; and the Space Telescope Science Institute will operate JWST after launch.

Conclusions

We are privileged to live in an age of unprecedented exploration. In the four centuries since the telescope was invented, our knowledge and our view of the physical universe has grown by leaps and bounds. We have found two new major worlds, swarms of asteroids, lots of small worlds beyond Neptune, a great number of deep sky objects, hundreds of billions of galaxies, and an expanding universe. For the first time in history we have sent machines beyond the confines of Earth into the depths of space, with the farthest man-made objects (*Voyagers 1* and *2*) continuing to move deeper and deeper into space at the rate of 35,000 and 39,000 miles per hour. As of March 30, 2012 (16:30UT). *Voyager 1* is 17,932,000,000 km (119.9 AU) from the Sun and 17,983,000,000 km (120.2 AU) from Earth; *Voyager 2* is a bit closer, at 14,700,000,000 km (98.3 AU) from Earth and 14,678,000,000 km (98.1 AU) from the Sun. It would take over 33 h to send and receive a message from *Voyager 1,* and a little over 27 h to send and receive a message from *Voyager 2.* Both spacecraft are still operating at the edge of the Solar System.

After perfecting the art and science of telescope making on Earth, we have lofted several into space for unprecedented views of faint (and not so faint) objects. The best known of these space telescopes is the Hubble Space Telescope, which has sent back hundreds of thousands of images of object humankind were imaged for the first time ever. We have seen unprecedented and crystal clear views of nebulae, star clusters, supernova remnants, and galaxies, views that can only be matched by actually going there. The James Webb Space Telescope promises to exceed Hubble's accomplishments when it is launched in 2018 as it looks even deeper into the cosmos (Fig. 6.13).

Because of the finite speed of light, whenever we look at something in the sky we are looking at it as it appeared at some point in the past. For nearby objects such as the Sun, Moon, and planets, we are seeing light that has been emitted or reflected

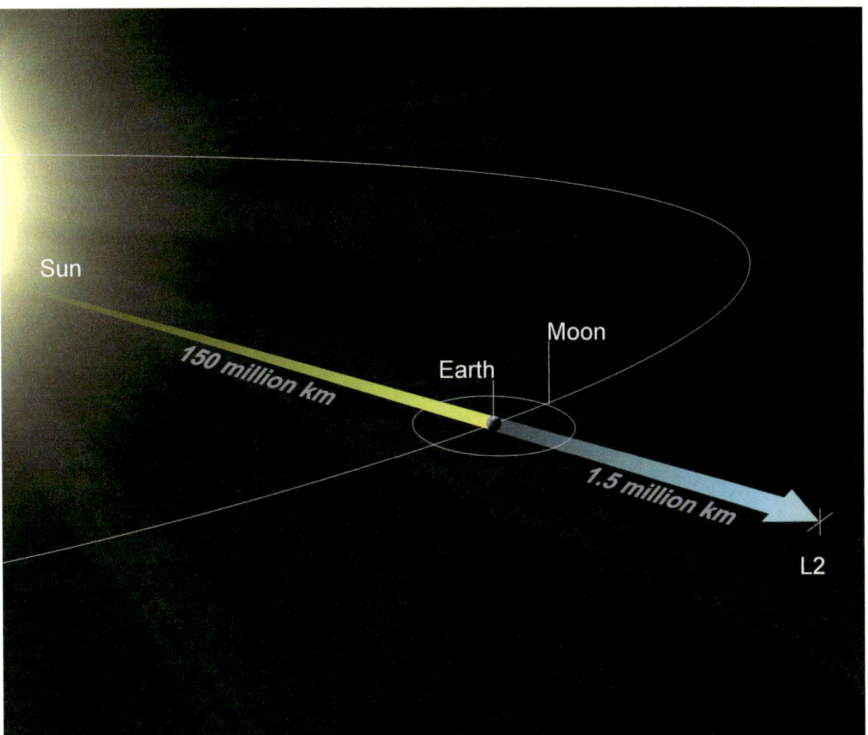

Fig. 6.13. Orbit of the James Webb Space Telescope, near the L2 point (Image courtesy of NASA).

from these objects and then traveled interplanetary space to reach us. We are seeing each as they were at some time in the past, from 1.3 s for the Moon to several hours for the outer planets. As we look at the stars in the nighttime sky, we are seeing them as they appear years, decades, even centuries ago, again since it takes light that amount of time to get from the source to our eyes and instruments. Faint objects give off less light, or their light is dimmed by some agent (dust or distance), so we are seeing them as being faint, but the same rules apply. For objects within our galaxy we are seeing these as they appear decades to millennia ago.

Outside the Milky Way Galaxy, we are looking even farther into the past as we look deeper into intergalactic space. The Magellanic Clouds are seen as they were some 185,000–200,000 years ago. The great Andromeda Galaxy greets us with 2.5-million-year-old light. Other bright galaxies show themselves to our eyes or detectors as they appeared 10, 20, up to 60 million years or more in the past. Fainter objects, with the light having traveled up to hundreds of millions of years, appear as they did that many years ago.

The most distant quasars visually accessible through backyard telescopes appear as they looked several billion years ago. These exotic objects are so far away that it took their light that long to reach our part of the universe. As we look at more and more distant objects we are looking more and more back in time. In fact, the history of the universe is laid out in the night sky as we observe objects at various distances from Earth and at various times in the history of the universe. With the proliferation of larger and larger telescopes, both ground- and space-based, we are seeing ever fainter and fainter objects at greater and greater distances in space and time. It should only be a matter of time when we will be able to glimpse the ultimate faint objects…the very first galaxies to populate the very young universe. Perhaps this will come with the launch and establishment of the JWST.

So the next time you are out looking for the faintest of objects, remember that the seemingly very few photons your eyes receive from these faint objects have been traveling intergalactic and interstellar space almost undisturbed since the distant past. You are seeing these objects as they looked a long time ago. This fact is yet another reason why deep astronomy is such a fascinating hobby to pursue.

Part II

How to Observe Faint Objects

Chapter 7

General Guidelines for Observing Faint Objects

Introduction

This chapter introduces some of the general guidelines for starting an observing program for faint objects. The objects featured in this book are mostly those that need a telescope with an aperture of 10 in. (25.4 cm) or more, observing in relatively dark to very dark skies. One may also use these techniques to observe somewhat brighter objects from light-polluted locations or with smaller instruments. What is common for both setups is that one is looking for something that lies at or near the edge of visibility and that person is trying to get the most out of the observation.

Let's assume the type of 'scope used in carrying out these observations is the reflector or Cassegrain variety, with clean optics and clean eyepieces. Although the minimum size aperture needed to observe the objects that will be presented later is 10 in. (25.4 cm), some of the brightest objects can easily be seen through an 8 in. (20.3 cm) under ideal, dark skies. Most of the objects will be accessible with 'scopes in the 10–12 in. (25.4–30.5 cm) range, but some of them will need larger instruments to be seen, even under dark, transparent skies. As far as the basics of how to use a telescope, the reader should consult other literature that has such information. The same concerning the basics about cleaning and collimating; again, here the focus is on what to do once one gets to the field and is ready to observe. Useful information in this regard will be provided in Appendix B of this book, as well as planning prior to the observing session.

What Is Needed to See Faint Objects

The object here are faint objects, observed with larger 'scopes from relatively dark to very dark skies. Aside from the minimum aperture already mentioned, another necessity is that the telescope has clean optics that are precisely collimated. If there is any dirt or misalignment, these will smear out what little light you get from your faint object, which could make the difference between something that is just visible and something that is invisible. A thorough familiarity with your equipment is a must, especially with the GOTO part if your telescope is equipped with one. If you do not have a clock drive or GOTO (and even if you do), there are some additional resources that you will need either printed (e.g., Uranometria star charts) or digital

B. Cudnik, *Faint Objects and How to Observe Them*, Astronomers' Observing Guides,
DOI 10.1007/978-1-4419-6757-2_7, © Springer Science+Business Media New York 2013

(e.g., planetarium software). You will also need to be intimately familiar with these resources to get the best use out of them.

You will also need dark, moonless, and clear skies and utilities to predict when the skies will be clear and dark. We go into more detail about these in the next chapter. Along with the skies and the 'scope, you will likely need special filters to help tease out the faint object from the faint background sky glow. These filters include the OIII filter, the UHC (Ultra High Contrast) filter, the LPR (Light Pollution Rejection) filter, and more. Also, soon after arriving at your site location, you will need dark adaptation and a means to preserve this. You will need to have patience and develop experience as you hunt these objects, which are not for the faint of heart (pun intended).

Experience is not a strict requirement, as you have to start somewhere, but it does help in locating your objects. As you set about finding them, basic to intermediate knowledge of the night sky (where major constellations and stars are at for the time of year you are observing, as well as the nightly movement of the sky) and star hopping skills will be needed, the latter especially if you do not have GOTO capabilities with your 'scope.

It is also a good idea to have reasonable expectations on what you will actually see. Don't rule out disappointment, as it will happen, but remember that losses do not keep your favorite sports team (if you are into sports, that is…) from continuing their pursuit of the championship. You will be hunting for objects that may not show up in your eyepiece at first, but as you look at more and more objects under different conditions, you will be able to judge more effectively what is accessible and what is not with your setup.

With this in mind, there are two considerations to remember as you go after faint objects: the overall magnitude of the object and its surface brightness. Depending on what type of object you go after, this will require different techniques, telescopes, and skies to see. Galaxies tend to have relatively high surface brightnesses near their core, but nebulae are a different story. (This issue was addressed briefly at the end of Part I while talking about some of the reasons behind the faintness of faint objects.)

A Faint Object Example: Planetary Nebulae

Some planetary nebulae appear point-like, with high surface brightness, but their point-like appearance means they blend in quite nicely with the surrounding star field, especially if the star field happens to be crowded. An OIII filter helps to pick out the planetary nebula from the surrounding stars. It blocks all light except for a special shade of green at 500.7 nm, the OIII (this symbolizes oxygen with two electrons removed) emission line. Since planetary nebulae emit much of their visible light in the OIII emission line, which penetrates through the filter almost undimmed, these stay the same brightness while the surrounding stars are dimmed considerably by the filter, since most of the light from these stars is blocked by the filter.

By passing an OIII filter in and out of your view as you look through the eyepiece (a process called "blinking"), you will be able to easily tell which one is the planetary nebula. The stars will dim, but the nebula stays the same, which makes

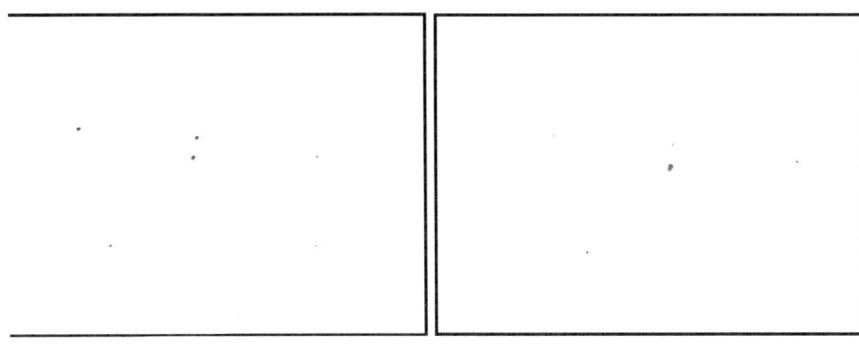

Fig. 7.1. Author's drawing of the effect of an OIII filter on a stellar planetary nebula (Vyssotsky 1–4 = PK 27–3.2), without the filter (*left*) and with the filter (*right*).

it stand out in the view with the OIII filter in place. Note the flattened diamond-shaped configuration of stars in the left half of Fig. 7.1. The bottom star of the diamond is the planetary nebula. The right half of the figure is the same view, but through an OIII filter. Notice that the planetary nebula stays the same (almost), but the rest of the diamond has nearly faded nearly. This nicely shows the contrast increase that is provided by the OIII filter for certain objects, including planetaries.

Other planetary nebulae are a lot more subtle. They have such a large apparent diameter that they are spread out over a huge area, making their surface brightnesses very low. In some cases, an OIII filter becomes an absolute necessity in order to even *see* the object, and even then it can be hard to pick out from the background sky glow, especially if there is stray light pollution, moonlight, haze, or mist present. With such types of objects, a larger, wide-field telescope is useful, but one can achieve a positive observation with a slightly smaller 'scope at higher powers, used under the darkest and most transparent skies possible.

The Level of Observational Challenge

One may be able to gauge the challenge of a faint object by designating the type of equipment an observer of average ability would need to see the object with certainty. Since this varies widely from observer to observer, site to site, and instrument to instrument, we will not attempt to categorize the objects presented in Chap. 10 by challenge level, but we will bring to the reader's attention that one can assign a skill level to each object. In general there is a minimum aperture recommended for observing various objects at a dark sky that has a zenith limiting visual magnitude of 6.5. One may want to experiment with one's equipment to determine its limits and share with others. As mentioned previously the focus is primarily on objects that can be seen with a minimum aperture of 10 in., but there are objects within the lists presented in Chap. 10 that need larger telescopes to see visually.

One can group the objects into three tiers or skill levels. Each is summarized below. Note that this is not an authoritative classification scheme; rather, it is a suggested guideline that one can use to gauge what one can see through one's telescope.

It may be possible to see some Tier II objects with 'scopes of 14–16-in. (35.6–40.6-cm) apertures; although an 18-in. (45.7-cm) is recommended. The three tier levels of skill are summarized:

- TIER I: The primary focus of this book and the majority of objects presented; these are objects that need a 10–18-in. (25.4–45.7-cm) 'scope in order to observe them.
- TIER II: Objects that need 18–30-in. (45.7–76.2-cm) 'scope to be observed and area secondary focus of this book (since fewer people own 'scopes larger than 18 in. in aperture).
- TIER III: These are objects that need at least a 30-in. 'scope to see them. This group of extremely faint objects is a tertiary focus for this book, since very few people own such large telescopes.

Getting Started

To best prepare for a productive observing run, here are some guidelines to keep in mind (these are presented in the form of questions that you can ask yourself as you prepare). Each is explained in more detail in the next chapter:

- What type of telescope and equipment do I have?
- What are the capabilities of my equipment?
- What is my own level of experience?
- What type of star atlas or astronomical software do I have or can get that will show faint enough stars and other objects to enable positive identification of them?
- What quality of skies do I have access to? Can I get to a reasonably dark sky in a reasonable amount of time or do I need to go far (and if I go far, do I plan to camp a few nights at the site)?
- What do I want to see? Galaxies? Star clusters? Galaxy clusters? Nebulae? All of these?
- How many of these objects do I want to see? How much time do I have to see these in? Do I want to try for as many as possible, or can I take it slow and enjoy the view?
- Do I want to do this leisurely? Unstructured? Or do I want to work through a list of objects? What is the total number of objects that I want to see for this project?
- For larger 'scopes that need to be set up, how much time do I need to set these up?
- Do I have a checklist to make sure I have everything I need (next chapter)?
- What is my weather criterion for attempting an observing run? Do I only go under the best possible conditions? Or can I make do with a mediocre night of transparency?
- What kind of filter do I need? As mentioned, there are several kinds of filters on the market that improve the view of objects of interest.

When engaging in any form of advanced astronomy, there will need to be at least some form of preparation, and the above checklist is a good resource to start with. All of the bullet points may not necessarily apply to you, but you can make use of as many (or as few) of those as you need. The above text is a more general activity checklist, meaning there are things to consider before embarking on a project to observe faint objects visually. More specific checklists, at least suggested ones, are provided in the next chapters and these apply to individual observing runs themselves.

One final question: how do I know if I actually see it? There are several ways one can verify for oneself if one has seen the object or not (% of time observing the object when it is actually visible, verification with an observing buddy, etc.). Here is the criterion that you might use: Try to spot it on two to three separate occasions (each occasion being at least a half second in duration), separated by a minute or more each, during a single observing session. If you spot the same smudge in the same place on all of these occasions, it is more likely this is the actual object and not merely your imagination. Confirmation by an equally skilled or more skilled observer would certainly help out, and confirmation using someone else's larger telescope aimed at the same object would make your observation almost certain (be aware that it is possible you could still mentally "project" the remembered image from the larger 'scope onto the star field revealed in your smaller 'scope).

A recent (at the time of this writing) article in an astronomically themed magazine presented the question as to how one can know what to look for prior to looking for challenging objects? One aspect of the answer was the fact that it is helpful not to have overarching expectations of what you will see. This is helpful to an extent, but if you are just starting out on the observation of faint objects, you will need to actually spend time in the field searching, finding, and even not finding objects. Experience is an excellent instructor as to what to expect, and the more you get out and observe, the more objects you see (and don't see), the more you will know what to look for.

This can still be a struggle from time to time in the area of comets. Sometimes you can get faint comets, but more often than not, you do not. After nearly 29 years of observing comets, bright and faint, this author has accumulated some experience to know what to look for, what limitations you have with a given comet, 'scope, sky conditions, etc., but since comets are dynamic, oftentimes they do not appear when you expect them to be easy targets, but (more rarely) they appear even if they were not expected to do so. The faint objects focused on in this book are a bit more stable and predictable in brightness. Since they are (for our intents and purposes) static objects (although dynamic in space scales smaller than what we can usually see and on time scales longer than we are able to live to see) their brightness is constant. Thus their appearance is affected "only" by instrument, sky conditions, and observer experience. The "dynamic" parts of this observing, perhaps even unpredictable parts, are the sky conditions. A small change in transparency can render an otherwise visible faint object invisible, and this can change in a single night. As already said, the more experience you can pick up (potentially) the more faint objects you can pick up.

Conclusions

As we begin our journey into the fainter parts of the universe, we saw that preparation is necessary to get started. That preparation includes the equipment that one will use to scope out these objects, the site from which these objects will be observed, and the observing knowledge and skill that will maximize the chances of finding faint objects. An overview of these things is above, and we will go into considerably more detail in Chap. 8. This chapter will give information on how to find a safe, suitable dark sky observing site; the best way to prepare the telescope and accessories; planning and executing an observing run (and how to get the most out of the time spent observing); and considering how a particular run went after the fact, in order to improve the next run. In order to help with the planning and preparation, we present a little faint object observation theory to take into account.

Chapter 9 provides information on the observing lists that are laid out in the chapter that follows it; it also shows where the interested observer may find suggested beginner observing lists if he or she is just starting out observing faint objects. Next, the chapter introduces a broad sampling of faint objects collected from a variety of sources, that the observer can delve into. These objects are presented in more detail in tabular form in Chap. 10. Chapter 11 provides more detail on what to record and how the observer can stick with their observing programs, and Chap. 12 also provides the scientific side of faint object observing, to include a brief tutorial on imaging. With this, a handful of projects are featured that are scientifically useful for visual astronomers to undertake.

In conclusion the first step to a viable, rewarding observing program is to make a plan based on available resources, or planned resource acquisition, and determine what kind of and how many objects you want to observe. Plan out how many times per month or year that you can go out to observe and work on your projects. Also determine what accessories you will need to bring with you such as eyepieces, filters, charts and others. As just mentioned, we will go into more detail about these things in the following chapters.

Chapter 8

Preparation and the Observing Session

Choosing a Dark Sky Observing Site

One of the essentials of faint object observing is a dark sky observing site [90]. This can be a site that is owned or leased by your local astronomy club. The author uses a location that is simply referred to as "The Dark Site," an 18 acre piece of land some 7 miles west of Columbus, Texas, which is managed by the Houston Astronomical Society for the use of its members and guests (a family who is well known in the greater Houston area owns the property as of this writing). Other clubs may or may not have dark sites of their own. The preferred site is one that is owned or operated by someone you know (and you have permission to use their land), or is a site where a "star party" is held annually or more frequently. Annual star party events such as the Texas Star Party are excellent venues for observing faint objects. Don't just pick an arbitrary field without knowing who it belongs to or what it is – remember, SAFETY FIRST.

An excellent way to find a dark site is to join an astronomy club that has access to a dark site, go through their required training, and you're there! Or you can search for the nearest star parties that happen throughout the year and attend them. Alternately, if a friend of yours has a property at a dark sky location, you can, with your friend's permission, use it. Between these options, you are most certainly going to be able to find a safe and useful site where you can view your faint objects.

Make sure that the site you pick is one that has a limiting magnitude of 6.0 or fainter. If you live in the United States, Canada, northern Mexico, or the Bahamas, you can make use of a resource that predicts different aspects of sky conditions for astronomical purposes. You can likely find your site, or one near your site, by going to www.cleardarksky.com (Clear Sky Charts),[1] select the state or province you live in, and look at the color-coded square to the right of the clear sky prediction chart. This is under a column headed "light pollution" and represents the Bortle dark sky class for that site. Select a site that has a Bortle class of between 1 (darkest) and 4 (somewhat dark); skies with Bortle class 5 and up become too bright for the faintest (and those with low surface brightness) objects. The Bortle class is olor-coded

[1] As is the case throughout this book, websites are frequently referred to. Over time it is likely that such websites will change web addresses or simply go out of existence. This is something for the reader to be aware of as time passes.

B. Cudnik, *Faint Objects and How to Observe Them*, Astronomers' Observing Guides, DOI 10.1007/978-1-4419-6757-2_8, © Springer Science+Business Media New York 2013

Table 8.1. Bortle's classification scheme of sky darkness [92]

Class	Title	Color key	Naked eye L.M.	Stellar L.M. with 12.5D reflector	Description
1	Excellent dark-sky site	black	7.6–8.0	19 at best	Zodiacal light, gegenschein, zodiacal band visible; M33 direct vision naked-eye object; Scorpius and Sagittarius regions of the Milky Way cast obvious shadows on the ground; airglow is readily visible; Jupiter and Venus affect dark adaptation; surroundings basically invisible
2	Typical truly dark site	Gray	7.1–7.5	17 at best	Airglow weakly visible near horizon; M33 easily seen with naked eye; highly structured summer Milky Way; distinctly yellowish zodiacal light bright enough to cast shadows at dusk and dawn; clouds only visible as dark holes; surroundings still only barely visible silhouetted against the sky; many Messier globular clusters still distinct naked-eye objects
3	Rural sky	Blue	6.6–7.0	16 at best	Some light pollution evident at the horizon; clouds illuminated near horizon, dark overhead; Milky Way still appears complex; M15, M4, M5, and M22 distinct naked-eye objects; M33 easily visible with averted vision; zodiacal light striking in spring and autumn, color still visible; nearer surroundings vaguely visible
4	Rural/suburban transition	Green/yellow	6.1–6.5	15.5 at best	Light pollution domes visible in various directions over the horizon; zodiacal light is still visible, but not even halfway extending to the zenith at dusk or dawn; Milky Way above the horizon still impressive, but lacks most of the finer details; M33 a difficult averted vision object, only visible when higher than 55°; clouds illuminated in the directions of the light sources, but still dark overhead; surroundings clearly visible, even at a distance
5	Suburban sky	Orange	5.6–6.0	15 at best	Only hints of zodiacal light are seen on the best nights in autumn and spring; Milky Way is very weak or invisible near the horizon and looks washed out overhead; light sources visible in most, if not all, directions; clouds are noticeably brighter than the sky
6	Bright suburban sky	Red	5.1–5.5	14.5 at best	Zodiacal light is invisible; Milky Way only visible near the zenith; sky within 35° from the horizon glows grayish white; clouds anywhere in the sky appear fairly bright; surroundings easily visible; M33 is impossible to see without at least binoculars, M31 is modestly apparent to the unaided eye
7	Suburban/urban transition or Full Moon	Read	4.6–5.0	14 at best	Entire sky has a grayish-white hue; strong light sources evident in all directions; Milky Way invisible; M31 and M44 may be glimpsed with the naked eye, but are very indistinct; clouds are brightly lit; even in moderate-sized telescopes the brightest Messier objects are only ghosts of their true selves At a full moon night the sky is not better than this rating even at the darkest locations with the difference that the sky appears more blue than orangish white at otherwise dark locations
8	City sky	White	4.1–4.5	13.5 at best	Sky glows white or orange – one can easily read; M31 and M44 are barely glimpsed by an experienced observer on good nights; even with telescope, only bright Messier objects can be detected; stars forming familiar constellation patterns may be weak or completely invisible
9	Inner-city sky	white	4.0 at best	13 at best	Sky is brilliantly lit, with many stars forming constellations invisible and many weaker constellations invisible; aside from Pleiades, no Messier object is visible to the naked eye; only objects to provide fairly pleasant views are the Moon, the planets, and a few of the brightest star clusters

like so: black = 1, gray = 2, blue = 3, green = 4. Cities have skies with Bortle classes of 8 or 9, and the suburbs, depending how far from city center they are located, and depending on how many light sources they have themselves, have Bortle class skies of 5, 6, or 7 (Table 8.1).

The best observing sites have no stray light and air pollution, have a calm and dry atmosphere, have mainly clear nights, are not too far away (1 h or less drive time), are easy to access, and are safe. Such sites are hard to find, so you will probably have to make do with what you can find. This author was fortunate enough in the mid 1990s to have access to a 21-in. (53.3-cm) Cassegrain-type telescope at the Mount Laguna Observatory in California, a truly dark-sky site that was only a 50 min drive from my (then) home in San Diego. Some people are fortunate enough to be able to retire and live in or very near the darkest skies of west Texas or New Mexico, which is also a plus. But most people have access, assuming transportation is not an issue, to reasonably dark skies within a reasonable driving distance.

Atmospheric Conditions

On-site testing is needed to determine the quality of the site. Inspect the terrain during the daylight for rocks and other hazards. Daytime heating, leading to radiational cooling at night, will spoil the local seeing. Keep away from wooded areas: wind blowing over trees causes turbulence. High places are best because they are often above the inversion layer in autumn and winter, which leads to relatively warm, dark, dry nights. Better still are islands with high mountains. In these cases, there is stable inversion over long periods. La Palma, Hawaii, is one such example (Table 8.2).

The main drawback of these high sites is a low oxygen level. This lack of oxygen results in loss of dark adaptability, which leads to the inability to see faint objects. If you have the opportunity to observe at such location, allow several days for dark (and altitude) adaptation.

Don't observe alone unless you are very familiar with the site. If you are stressed or nervous or otherwise anxious, that reduces perception, but bringing your favorite "star gazing" music does help. Group settings can be more pleasant but distracting if you are a serious observer. Be relaxed physically and mentally at the eyepiece. Also, limit alcohol and nicotine intake, since these reduce visual acuity and dark adaptation.

Weather Conditions

Obviously clear moonless skies (or skies with only a thin crescent Moon) are needed, and here is where the Clear Sky Charts become useful. Select nights around new Moon, when the weather is expected to be clear (with no fog, no high

Table 8.2. Antoniadi's scale of seeing quality [93].

Level	Seeing	Image
I	Perfect	Calm image without a quiver
II	Good	Calm images over several seconds, with brief interruptions of slight quivering
III	Moderate	Large air tremors that blur the image
IV	Poor	Constant troublesome undulations of the image
V	Very bad	Image is hardly enough to allow a rough sketch to be made

cirrus clouds, and no haze or mist) AND transparent. To get the most out of your observing, go for the best conditions available. You can also make use of substandard conditions to an extent to view the brightest of the faint objects, but you can also use the dark half of the quarter Moon night (which is after moonset or after local midnight for the first quarter, or before moonrise/local midnight for the last quarter) as well. Unfavorable conditions may crop up during the night, after what began as an apparently clear and transparent evening, so in the event of the unexpected, it is good to be flexible to make the most of what you get.

In summary, points to remember with regards to weather conditions

- Weather resources to help determine the quality of a prospective observing run include clear sky charts, cloud forecasts for astronomical purposes, "skippy" sky.
- Skies need to be moonless or mostly so – either new Moon or less illumination than a quarter phase; even with this phase, focus on dim objects only during the moonless part of the night.
- Skies should be free of high cirrus clouds, haze, excessive moisture (high humidity makes for poor transparency and condensation on telescope optics).
- If the weather turns out to be less than ideal, move to brighter objects or less challenging ones.

Maintenance and Preparation of the Telescope

Depending on your telescope, vehicle, driving distance, number of nights you will spend, etc., will determine how you prepare for your observing run. It is best to have a written checklist (speaking from much experience!) of every little item prepared ahead of time; work your way systematically through the list, from top to bottom, to ensure you have everything you need. Double and triple check that you have everything you need prior to departure.

As you prepare your 'scope and accessories for observing, there are considerations regarding finderscopes, eyepieces, and filters. We will leave the basics to other books but rather focus on items that pertain to the observations of faint objects (primarily). What will follow are some things to consider when selecting accessories for your observing program.

Eyepieces

High quality (but don't "go overboard" in terms of price), which includes sharp images up to high magnification, high contrast, and large apparent field of view. Although it varies from 'scope to 'scope, an eyepiece with a 70–80° field of view and large eye relief (how far you can place your eye from the eyepiece and still get a complete field of view) is recommended to allow people with eyeglasses to view. It is also a useful thing to have a rubber eye guard or cover your head and eyepiece with a black cloth to keep stray light out and allow you to see fainter objects.

Table 8.3. Filters useful for deep sky observing

Filter and brand name example	Bandpass	Comments
Narrowband filter (e.g.: Lumicon Ultra High Contrast or UHC)	22–26 nm	It works by isolating the two doubly-ionized oxygen emission lines (OIII, 496 and 501 nm) and the hydrogen-beta line (486 nm) and blocking the rest. Useful in dark sky and light-polluted settings and is useful for bringing out more detail in extended emission nebulae as viewed from dark sky observing sites
Broadband filter (Lumicon Deepsky filter)	90 nm	Works by blocking light pollution (mercury vapor, high and low pressure sodium vapor) from street lights, neon lights, and airglow while letting the light from nebulae pass through. Best visual light pollution filter for improved views from urban areas
OIII Line Filter	10–12 nm	Best used with planetary nebula and is the filter of choice for "blinking", that is to pass the filter in and out of the visual line of sight to pick out a stellar or small planetary nebula from a star field. Isolates only the two OIII lines emitted by these objects. Useful for increasing the contrast and bringing out detail in many emission nebulae also
H-Beta Line Filter	8–10 nm	Isolates the H-beta line alone (486 nm) and is excellent for viewing faint emission nebulae. In fact certain nebulae can only be viewed with this filter, which is best used under clear, transparent skies with larger scopes

More information about these filters can be found at the Lumicon website [94].

Filters

Several types are available (Table 8.3). Most popular types include the LPR (Light Pollution Rejection) filters that help to reduce the glow of light from artificial sources, but allows wide bandpasses of light through OIII (495.9 and 500.7 nm) and H-alpha (656.3 nm). H-beta (489.9 nm) are ultra-high contrast (UHC) filters that enhance contrast of HII regions and planetary nebulae. The filters allow bandpasses of 30–50 nm. A "line filter" shows a band pass of 10–15 nm centered on a single spectral line (e.g. OIII). Filters come in two standard sizes: 1–1/4 and 2 in. (31.8 and 50.8 mm), and the 2-in. is more than twice the price of an 1–1/4 in. version.

Finderscopes

Telrad makes a good starting point to get to the right area of the sky, but you will need a high quality finderscope to fine-tune your aim. It is vital that the finderscope is correctly aligned and focused before beginning an observing session. You can either use a star that moves with Earth's rotation. Or you can align with a terrestrial object by day before it gets too dark. The optimal size for a finderscope is 40–60 mm aperture with a 2–5° field of view. A magnification of 10× is good for a 50 mm lens.

Getting to the Celestial Target

GOTO is quite useful, but there are things to keep in mind. Knowledge of the sky is helpful, but sky computers and planetarium programs typically result in no practical experience at learning stars and constellations. If you are unfamiliar with the sky, take time to learn the layout of the constellations and where the bright stars are located. Also make it a point to learn star hopping; if the object you are looking for is quite faint and you are going by a star map or photograph, you will need to learn how to match the stars you see in the eyepiece with that you see in the image. Two things can trip you up during the final steps of finding the object: the size/resolution of the eyepiece's FOV versus the image of map or image, and the faintest star visible in each. Here is what to do:

- Identify the two or three brighter stars that make a pattern that you can also identify on the chart or image. Ensure you are using the correct orientation of the image to match the eyepiece (e.g., south up, mirror or straight view, etc.)
- Locate the object of interest on the image or map and plot a "visual hike" from your "beacons" to the object. Ensure that the stars you pick to "hop" are also visible in the eyepiece. You may need to do several iterations of looking at the charts, eyepiece, charts, etc.
- Try to "frame" the object's position in a pattern of three or four surrounding stars, first on the image, then through the eyepiece. Mentally mark where the image is in the eyepiece FOV, then use averted vision, filters, etc., to try to pick out the object.

When you are "going after" an object (as already discussed) you may need to do some star hopping, depending on the type of instrument you are using. We will assume the reader (you) has experience at star hopping and will refer you to works such as "Galaxies and How to Observe Them," which provides a more detailed description of star hopping, the theory of visual observations, and other topics briefly touched upon in this chapter. The following is a review to help get you to your object. If you have a Telrad you can use it to aim your telescope to as close to the object's position as possible, or aim it at a star that you can use as a starting point to "hop off of." Next, using the lowest magnification/widest field of view eyepiece in your kit, ensure you either have the object's field of view in the eyepiece or you can get to it in fairly short order. Carefully move the 'scope from star to star as you "close in" on your object's location. Upon arrival, your faint object may not be visible right away, so you will need to change magnifications (if appropriate) and/or use other techniques we have discussed so far, such as averted vision, field sweeping, filters and others.

The "Care and Feeding" of Your Telescope

No matter what kind of telescope you have (with some exceptions), if you transport it for trips longer than out the back door to your backyard, you will likely need to collimate or recollimate your 'scope at setup (unless the 'scope is permanently mounted at a fixed observatory). If this is a new 'scope, or a 'scope that you do not use often, practice setting up and tearing it down at home before attempting to do it in the field (where it may be dark at the time you do this; it's a good idea to arrive at the observing site at least 2 h before sunset).

Ensure that your mirrors and eyepieces are as clean as possible. You can blow dust off your mirrors with canned air, the kind used on computer keyboards to get the debris out. Take care not to touch the mirror with your bare hands. It is best not to clean the optics unless it becomes absolutely necessary, but over time loose debris will accumulate on the mirror. This is where the air comes in handy. If more cleaning is needed, and this is a reflecting telescope, remove the mirror first, but keep it mounted in the mirror cell if possible. Next, make a dilute solution (15–20 %), mixing a mild dishwashing liquid and distilled water (tap water from the city or from a well have tiny particles that can scratch). Take cotton balls and use them to swab the mirror with the cleaning solution, starting in the center and working your way outward in a slow spiraling motion. Next, rinse the mirror off with the distilled water, and carefully blot up any leftover water drops with a paper towel.

The Schmidt-Cassegrain telescope (SCT), with its sealed tube structure, is not easily cleaned. However, that very same design keeps the primary and secondary mirrors clean, since dust and humidity are sealed out. The front corrector plate of the SCT is exposed to the elements and will collect dust and the occasional smudge over time. Because the corrector plate has delicate multicoatings you will want to avoid cleaning it unless it is absolutely necessary. Like the mirror, you can use canned air to blow dust off the plate or use a very soft camel-hair brush to brush off the surface without scratching it. If you go the air route, do not use it for too long for any one application, as drops of liquid propellant or condensation may form and drip on the glass (same goes for the mirror).

For SCTs, a solution of six parts isopropyl alcohol and four parts distilled water should be used, as recommended by Celestron. Add a couple drops of liquid dishwashing soap per quart if desired. Meade Corporation recommends one part isopropyl alcohol and two parts distilled water with one drop of biodegradable liquid dishwashing soap per pint of solution.

You will need to have your telescope's mirrors re-aluminized every decade or so (depending on how often you use yours). One can find a company or individual who does this work online (some are listed at http://www.amateurtelescopemaker. com/mc.htm), and these typically charge $70 or more for a 10-in. or larger mirror. Have the secondary done while you are at it.

Eyepieces can be cleaned following the instructions provided above. One additional item that would be useful is the LensPen™ (Nikon, Celestron, and others endorse this product), which professional photographers use to clean expensive camera lenses. A LensPen has a soft brush on one end to remove dust, and a soft pad on the other end to remove smudges. This accessory can easily get into hard-to-reach bottom glass surfaces of eyepieces [91].

Proper storage of 'scope and eyepieces also contribute to their care. It is best to store them in a cool, dry place, with the 'scope pointing upright and covered securely. Store the eyepieces in their original containers, with separators to prevent them from clanking into each other. Add silica gel bags to the eyepiece case to keep out the humidity; this will prevent mold.

If you are observing at a site where electricity is available, have a hair dryer on hand with which you can use to remove any dew that develops on your optics. Otherwise you can use a heated, battery-powered dew shield, which helps in most cases (except when the dew is extremely heavy).

The website www.cloudynights.com contains a lot of resources that are useful for the amateur astronomer. It provides a forum for discussion of a wide variety of topics, and, as the website name indicates, it gives the astronomer something to do if the night turns out to be cloudy.

To sum up, here are some points to consider when maintaining and preparing your telescope for observations:

- The optics of the 'scope need to be as cleaned as often as possible.
- Have a way of keeping dew from forming.
- Keep elements collimated to concentrate the meager available light from faint objects and prevent it from smearing out, rendering the object invisible.

In most places, mosquitoes can be a problem, so you might want to use a mosquito repellent for observing comfort. Speaking from experience, few things can bring frustration to an observing experience like trying to sight in a star and not being successful at it while being harassed by mosquitoes. Eliminate that distraction with repellent, but remember to apply it far away from any optics (binoculars, your telescope, other people's telescopes), and make sure your hands are free of repellent (at least the palms and fingers) so as not to smear any on eyepieces or "sticky up" control paddles. Be sure that mosquito repellent and a simple first aid kit (in the event of unexpected injuries) is listed on your observing checklist.

Planning the Observing Sessions

Whether you use sophisticated observing programs such as *Sky Tools* or simply use pencil and paper, it is vital that you plan ahead to get the most out of your observing time. Do you want to get as many objects as possible in a given session? Make your list but be realistic, and be prepared to drop some of the objects if the observing session time runs too fast and your progress in checking off objects is too slow. Perhaps have a hierarchy of objects – a handful of "must see" followed by a batch of "if possible" and "as time allows…" But whatever you do, be flexible, be content with what you do get, and don't merely "check the boxes." Enjoy the view! This will likely be the difference between a stressful session and a satisfying one. You do want to make the most of the precious little dark sky time.

Here are some general planning tips:

- Know exactly what you want to observe.
- Select your lists of objects to observe.
- Determine which objects can be seen during your current session.
- Estimate how much time you will spend per object.
- Ensure the objects are as high in the sky as possible (to minimize the atmospheric extinction or dimming).
- Get any observing aids ready and maintain a checklist of what will be brought to an observing session.

How many nights will you be observing? One night? Two nights (weekend)? Five to seven (for a star party)? Obviously more nights mean more objects. Plan to observe bright objects such as planets and bright stars in a separate session (perhaps dedicate one full night to bright objects only) so as to not degrade your night vision. Finally, what type of objects do you want to view? Do you want to observe all faint objects or a combination of bright and faint objects?

For a single night, determine whether you will be able to use the entire night (dusk or dawn) or if you will need to nap (usually 3–4 h) because of a commitment the next morning. Also find out if the Moon will be in the sky for part of the night and, as the time gets closer, whether the skies will be transparent or merely average. These are factors that go into planning a night of observations. Consider designating several hour-long chunks for dim objects, and then observe brighter objects at other times.

For a three-night run, you might dedicate one of the nights to brighter objects and all-time favorites, then nights 2 and/or 3 (or parts of each) to faint objects. There is some flexibility built in if the weather turns bad for one (or more) nights, and you can work from lists that are a lot longer than what you possibly can observe in a night.

For a five to seven night run you can go "hog wild." Have lots of objects listed to observe. But pace yourself to ensure it remains an enjoyable time. It may be useful to start the run with the faint objects and leave the easier ones for later. Advantages to doing it this way are (1) if the weather turns bad later in the week, you will have done your faint object observing, and (2) You may get rather tired and worn out by halfway through, and tired eyes will not be as adept at picking out the faint fuzzies as fresh, rested eyes.

Observing Software

There is an assortment of planetarium programs on the market today to assist the astronomer in planning an observing session. One program called *TheSky 6* is very useful for controlling GOTO telescopes such as the 14-in. Schmidt Cassegrain. Information on people's favorite astronomical software, along with descriptions and where you can purchase each, is given in Appendix C at the end of this book. Several people from different astronomical groups were surveyed, it turned out that there is a wide variety of favorite software that people use. More detail on each of these software are provided in Appendix C, but here are their names, in no particular order: *Megastar,* for planning observing sessions (and more); *AstroNotes,* for recording observations; and *Cartes du Ciel,* which is freeware that can be used for observing planning and execution (actually finding the objects). Also included is *Stellarium,* which can be used to teach astronomy; *TheSky* software, which is useful for tracking down faint objects (and *Sky 6* at a dark site for telescope control and to find the precise field where a faint object is located); and *ASCOM,* also to control telescopes.

Making the Most of Your Viewing Time

You may be the social type, spending as much (or more) of you dark time chatting with astro-buddies as you are looking (or not looking) through the eyepiece. Or you may want to maximize your time, working (almost) continuously from sundown to sunup.

There is really no right or wrong way to do it, but whatever you do, make sure it is enjoyable and you enjoy your dark time in your own way. Here are some things that can ensure a successful observing run:

- Plan your run ahead of time so you know what you are going to do once you set up.
- As long as it is clear, work on your observing list, getting as many objects as possible, yet pacing yourself so as to avoid stress (as much as possible) and maximize the enjoyment factor.

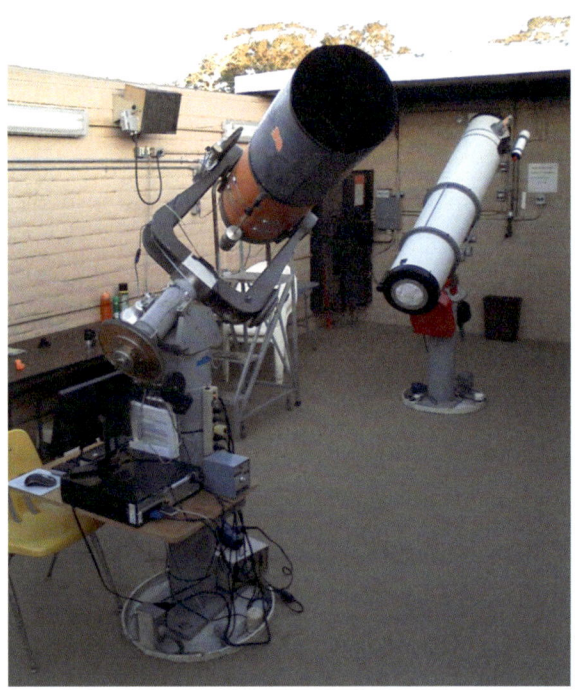

Fig. 8.1. The author's image of his dark site sanctuary – the Houston Astronomical Society observatory with the roof rolled off, showing the "C14" (14-in./35.6 cm f/11 Cassegraine in the foreground with its classic orange tube); and the "F7" (12.5-in./31.8 cm f/7 Newtonian reflector) further back.

- You may want to work without stopping, because dark time is a precious resource not to be wasted (especially if you live in light-polluted suburbia). Most of us live where light pollution reigns, and it takes time to plan and drive to a dark sky site (Fig. 8.1).

After your observing run, take time to reflect on your experience:

- How much were you able to get done?
- What were you not able to observe?
- How you would improve the experience for next time?
- What were some of the more memorable sights that you have seen?

No matter what you do it seems that the night goes by way too fast and the number of objects seen way too few by the time the dawn's early light begins to wash away faint objects. But you should feel satisfaction of having explored distant corners of the universe and having fun while doing so. The sense of accomplishment at what you were able to get, though you are exhausted due to lack of sleep, should leave you enthused at having completed another productive, meaningful, memorable night at the eyepiece.

Practical Use of Observing Theory and How the Eye Works

There are more ways we can use theory to enhance our observing (what we see, how to better see it or putting "theory" to use) [95]. Different people see things differently through the same telescope. Observing ability depends on experience and visual technique plus preparation on the part of the observer; atmospheric seeing, physiological state of the observer; and instrument quality on the part of the observing setup. In summary, here are the factors that go into determining the quality of observing for a given night:

- Sky conditions (transparency, seeing)
- Aperture
- Clean, collimated optics
- Entrance/exit pupil of eyepieces
- Dark adaptation
- Direct versus averted vision
- Contrast
- Magnification

The eye is extremely sensitive to motion and slight brightness differences. Cells that enable vision come in two varieties: rods and cones. Cones have their peak sensitivity at 550 nm and are primarily used during the daytime. They enable us to see color, but need a certain level of light to enable color to be perceived. Rods are useful in low light situations and perceive shades of gray only at a lower resolution. They are most sensitive ~20° around the fovea. Rods need 15–30 min to achieve their maximum sensitivity (let your eyes adjust for a minimum of 30 min before observing). In fact, preparations for a dark faint object session should begin 2–4 days in advance: avoid bright sunlight as much as possible, wear sunglasses and a hat. During observations, cover the head and eyepiece with a dark cloth. If you lose sensitivity, you lose your ability to see faint objects.

When using averted vision, avoid the blind spot. You can find the blind spot by looking directly at a star, and then shift your eye toward your nose until the star vanishes. Remember where you are looking during this instance to avoid inadvertently hitting your blind spot as you try to employ averted vision to glimpse a faint galaxy or nebula. The eye-brain system is capable of interesting things. The latitude of brightness is superior to video or imaging, as the eye can observe both bright and faint objects at the same time. The eye can even serve as an "image intensifier" – a time exposure on faint objects by using an aggregate of neural transmitters (ganglion cells) that combine several rod cells – effectively accumulating light like film but much less effective.

The "ease" at which one sees an image while looking through the eyepiece depends of the exit pupil of the eyepiece [96]. Many a beginner puts his or her eye to the eyepiece and sees only featureless black. After moving your head around "looking for" the image, it finally reveals itself, but either you have to have your eye right up to the eyepiece or hold your eye about an inch or two away from the eyepiece (by which time a lot of extraneous light comes in from the sides).

Fig. 8.2. Author's image of what the exit pupil looks like.

For the telescope, the entrance pupil is the aperture of the 'scope and is symbolized by P. For the eye, the entrance pupil is the iris opening. Although most of the diehard observers encountered are of the more mature variety, the physiology of the eye favors young people, who are typically inexperienced at astronomical observations. The maximum opening of the iris is from 6 to 8 mm, with older people having to make due with a 6 mm opening (but experience does make a good substitute for lack of ocular aperture, correct?) while the younger people have the luxury of a wider opening that lets in more photons that enables them to see faint objects more easily.

The exit pupil (Fig. 8.2) is defined as the amount of light coming from the 'scope and is abbreviated by p. In this case the size depends on the instrument's parameters only, namely the aperture ratio and the focal length of the telescope. So we have three cases that set up:

1. For $p < P$ all the light enters the eye from the eyepiece but there is room for more.
2. For $p > P$ only a fraction of the light is detected, with the rest lost.
3. For $p = P$ the situation is ideal, but only during maximum dark adaptation where $P_{max} = 7$ mm (for the average adult). But under these conditions, severe vignetting can occur, making it hard to see the image in the eyepiece.

Contrast and Magnification

A dark night sky has a brightness of 13.1 magnitudes per square arc minutes or 22.0 magnitudes per square arc seconds (these values are for the V_{mag}, for the B_{mag}; the sky is a magnitude fainter or 14.1 magnitudes per square arc minute). Contrast is defined as the ratio of the intensity of the signal to the intensity of the background noise (also known as "signal to noise ratio"). The human eye can effectively detect faint objects under low contrast. It is best to use a higher magnification with averted vision, which increases the apparent size of the object. The rod cells in the eye are responsible for "seeing in the dark," and a larger image involves more rods to collect the light from the image.

For magnification → magnification = telescope focal length/eyepiece focal length

If you know the eyepiece's apparent field of view, you can get the approximate actual field by taking the apparent field of view and dividing it by the magnification. To get a more accurate measure of the field of view for a given eyepiece, the following is suggested: Time how long it takes for a star to drift from edge to edge in your eyepiece. Ensure that the star is located near the celestial equator and that your 'scope is also facing south. Also make sure that the star crosses the center of the field of view. Once you get a good timing, take that timing, in seconds, and divide it by 120 s, and you will get your field of view in degrees.

Calculations by Mel Bartels (an American amateur astronomer and software development manager who has provided free software to control telescopes and is an experienced telescope mirror maker) show that the darkness of the night sky is more important than the aperture size of the telescope. Better a very dark sky with an 8-in. telescope than a light-polluted sky with an 18-in. telescope (an extreme example, but it shows the point…the larger telescope not only collects more photons from objects but also collects more photons from stray light pollution).

The visual observing experience depends on the condition of the atmosphere, which includes both seeing and transparency. John Bortle (famous for creating the light pollution scale, which is used as a standard reference for judging the quality of skies) has devised a scale of darkness that can be roughly quantified by the brightness of the sky. A Bortle 0.5 sky means 13.1 magnitudes per square arc second. In contrast, an urban sight (Bortle 7 or 8) is 12.1 magnitudes per square arc minutes. This increase in background sky brightness drastically limits the visibility of low surface-brightness objects.

Conclusions

Remember that any observing session needs pre- and post-processing to be fully meaningful. Also realize that no two observing sessions are going to be identical; there is usually something unique about each session. And just because you do not get to see everything on your list, appreciate the things you did get to see, and plan to pick up the missed objects on a subsequent observing session.

With all this said, one note bears repeating: once you acquire your object, take time to appreciate the beauty and the nature of what you are observing. The combined light from billons of unseen suns (if it is a galaxy), the immense distance/time that the light from the object has traveled to enter your 'scope and meet your eyeball, and the effort that you put into finding this object that likely few have seen. Realize that the object was probably unknown to humanity only a few short centuries ago and you will (very likely) be enjoying the object through optics that are far superior to those in the days of Charles Messier or William Herschel.

Chapter 9

Some Suggested Observing Projects

Introduction

There is literally a universe of objects competing (unbeknownst to the objects themselves, of course) for your attention at the eyepiece, and one question that you may be asking is: Where do I get started? This depends on what type(s) of object(s) you are interested in viewing and what your experience level is. Do you want to view only a small sampling of objects or do you like the challenge and satisfaction of long observing lists? What is the size and capability of your telescope, and what type of dark site do you have access to? These should all be taken into consideration when outlining an observing program.

Careful planning and having a procedure outlined prior to an observing run is important. Yet you also want to enjoy the freedom of spontaneity and not be confined or constrained to a single list, type of object, or process. So do a little of both! Make plans to observe a certain amount of faint objects, along with brighter objects and Solar System objects. Lay out the plan with enough "wiggle room" to change it if so desired, or if the weather changes unexpectedly. Sometime you can just go with little planning except to enjoy the beauty of the nighttime sky. Weather forecasts also influence the planning session (e.g., if clouds are expected to increase or decrease overnight, fog is expected late, etc.). This was addressed this earlier in the previous chapter.

Here, we will provide observing lists that you can choose from as well as a suggested progression of projects to follow. You will find a number of lists collected from several sources in Chap. 10, but the Astronomical League lists will need to be accessed through their website, which is in the next section. Also included is the online location of a group of lists published by various members of the Texas Star Party (TSP); the lists themselves are on the TSP website address, provided later in this chapter.

B. Cudnik, *Faint Objects and How to Observe Them*, Astronomers' Observing Guides, DOI 10.1007/978-1-4419-6757-2_9, © Springer Science+Business Media New York 2013

The Astronomical League Observing Programs

A great place to start your faint objects adventures is with the Astronomical League (AL). As of this writing, the AL boasts (at least) 37 observing programs called "clubs" that run the gamut of objects one can find in the universe, from the Solar System (and some events that take place in Earth's atmosphere, such as meteors and aurorae) to the most distant galaxy cluster or quasar. These can be accessed online at the following website: http://www.astroleague.org/al/obsclubs/ AlphabeticObservingClubs.html. As web sites do change over time, if the above site is no longer valid when you look, go to http://www.astroleague.org and look for the link to the observing clubs (the number of which are growing as new clubs are regularly added). Here are some suggested clubs to pursue, depending on your observing experience:

Observing Experience: Intermediate
   Caldwell Club
   Globular Cluster Club
   Herschel 400 Club

Observing Experience: Advanced
   Arp Peculiar Galaxy Club
   Galaxy Groups and Clusters Club
   Herschel II Club
   Local Galaxy Groups and Neighborhood Club
   Planetary Nebula Club

You may want to complete your requirements for Master Observer (you can find out what these requirements are at the AL website). If you do, you will have experienced many of these clubs, and have found that they are rewarding, indeed. You can start with the Herschel 400 Club, as there are lots of excellent targets on that list. Then do the Caldwell Club down to as far south in as you can go and still see the objects. The Caldwell list is very similar in nature to Messier's list, as the former, assembled by Sir Patrick Caldwell-Moore of England, included the best objects that were not listed in Messier's catalog.

The more advanced AL clubs offer objects that are more of a challenge to observe. Like any of the lists, there is a combination of easier and challenging targets to observe and enjoy, and these are excellent lists to get started in observing faint objects. The Arp Galaxy Club actually features four awards: two visual and two for imaging. One of the two (for each) consists of objects located in the northern celestial hemisphere, and the other features objects in the southern celestial hemisphere. For both the Arp and Galaxy Group clubs, you can select which 100 objects from each that you will observe, and within each list is a range of difficulties, from fairly easy to difficult. The Galaxy Groups objects are observable with a 12.5-in. telescope.

The Local Galaxy Groups and Neighborhood Club provide a list of very challenging objects, due to the fact that most Local Group galaxies are dwarf galaxies.

The description of this club states that all of the entries are within the range of 10–12.5-in. telescopes, but many of the individual groups of galaxies have objects beyond the reach of backyard telescopes. Finally the Planetary Nebula club consists of both bright objects and faint objects. Some of the challenges in this club include star-like planetary nebulae that reside in crowded star fields, larger low surface brightness objects, and small, faint proto-planetary nebulae.

Note that in order to receive a pin and certificate for completing a club, you have to be a member of the Astronomical League, either through a participating local astronomy club or directly with the AL. Please refer to their website for information on joining should you be interested in pursuing these awards.

Adventures in Deep Space

In addition to the clubs that the Astronomical League provides, there is an excellent online resource for challenging observing projects that you can use. It is called "Adventures in Deep Space: Challenging Observing Projects for Amateur Astronomers of All Ages." Among the lists provided in the next chapter, included is an especially challenging list of 75 objects to observe (the Challenge Sampler, Table 10.1), which were compiled by experienced deep space observer Steve Gottlieb for the Golden State Star Party. This list contains a broad sample of objects of varying levels of difficulty. Steve Gottlieb has regularly shared his experience with *Sky & Telescope* magazine and other sources.

Two more lists that are included in this book are also from the Adventures in Deep Space: the 75 brightest globular clusters of M31 (Table 10.6) and some of the brightest deep space objects that reside in M33 (Table 10.8). The latter is provided in tabular form along with an image that shows the locations of these objects in that galaxy, and included are Gottlieb's brief but informative descriptions as well. The Hickson Galaxy groups listing (Table 10.15) can be found also in Chap. 10 and the Arp catalog of galaxies (Table 10.2), as well as a listing of various objects that this author has assembled from various sources (Tables 10.2 and 10.14). Also included are the Palomar globular clusters along with other obscure globulars (Table 10.5).

The Adventures in Deep Space website (to access their lists, go to http://astronomy-mall.com/Adventures.In.Deep.Space/, and if this does not work do a web search for "Adventures in Deep Space" using a search engine of your choice) contains lists of objects on their catalog page to include the list below. In this book, you can also go to Appendix E for web addresses and more information. Here are the names of most of the observing projects that you will find on the "Adventures of Deep Space" website:

- 100 Peculiar Galaxies
- Off the Beaten Path (objects organized by season)
- Observing Galaxy Clusters (there are links to images of 16 clusters and superclusters of galaxies a handful of which are provided in the next chapter)
- The Abell Planetaries (86 gems to test your observing skills)
- The Ultimate Observing List

- Steve Gottlieb's NGC Notes
- Catalogs, Lists, and Links to:

Abell Galaxy Clusters

GSSP General List (summer-html)

Lynds Dark Nebulae

Abell Planetary Nebulae

Herschel 400-I

M31 Globular Catalog

Arp Catalog

Herschel 400-II

M33 HII Regions and Star Clouds

Barnard Dark Objects

Herschel 2500

Off the Deep End (10 challenges)

Orion Deep Map 600 (Megastar file)

Hickson Galaxy Clusters

Shakhbazian Galaxy Groups

Galaxy Trios

Lynds Bright Nebulae

Sharpless HII Regions

GSSP Challenge List (summer-html)

- Detailed monthly observing lists from February 2007 to January 2010
- Observing Reports

And much more.

For "Steve Gottlieb's NGC Notes," he has documented observations and information on 7,177 NGC objects and an additional 750 IC objects. Next to the New General Catalogue itself, Gottlieb's may be the most complete listings with descriptions of NGC objects, and this brings to mind the work of Sir William Herschel, who systematically observed thousands of objects in his day. Another long term project that one may take on is to observe as many of the 7,840 NGC objects as possible. One can access data on these objects through the above Adventures website or from the NGC Project itself (online it can be accessed at http://www.ngcicproject.org/).

Other Lists

The short answer to where one can find more observing lists and other information is the previous section, Chap. 10, and Appendix E of this book. The lists described so far are certainly not the only ones available, but they give a substantial sampling of faint objects to get interested readers started in the exploration of objects near the edge of telescopic visibility. No doubt there are many other individuals, groups, astronomy clubs, and more that have compiled their own challenge lists for themselves, their membership, or the general public. Indeed, there has been a wide variety of challenge lists that have been produced over the years.

One such group of lists includes the advanced observing program of the Texas Star Party, one of the more famous of the star parties in North America that is held annually in west Texas. These observing lists can be found at http://texasstarparty. org/ and feature such challenges as super thin galaxies, the Local Group of galaxies, the "GD Planetary Nebulae" list, and "Explosions over TSP" list. The last listing includes supernovae remnants, starburst galaxies, Seyfert galaxies, quasars, and BL Lacertae objects. A new list is generated each year for the annual star party, so keep checking back for more observing challenges. This author has completed the "GD Planetary Nebulae" observing roster using a 10-in. f/6 Dobsonian reflector under the dark skies of the 2003 Texas Star Party and has included sketches of several of these objects above (Fig. 9.1).

In addition, a link and brief summary (for southern hemisphere observers) of a project to observe the nebulae and clusters of the Magellanic Clouds is provided in

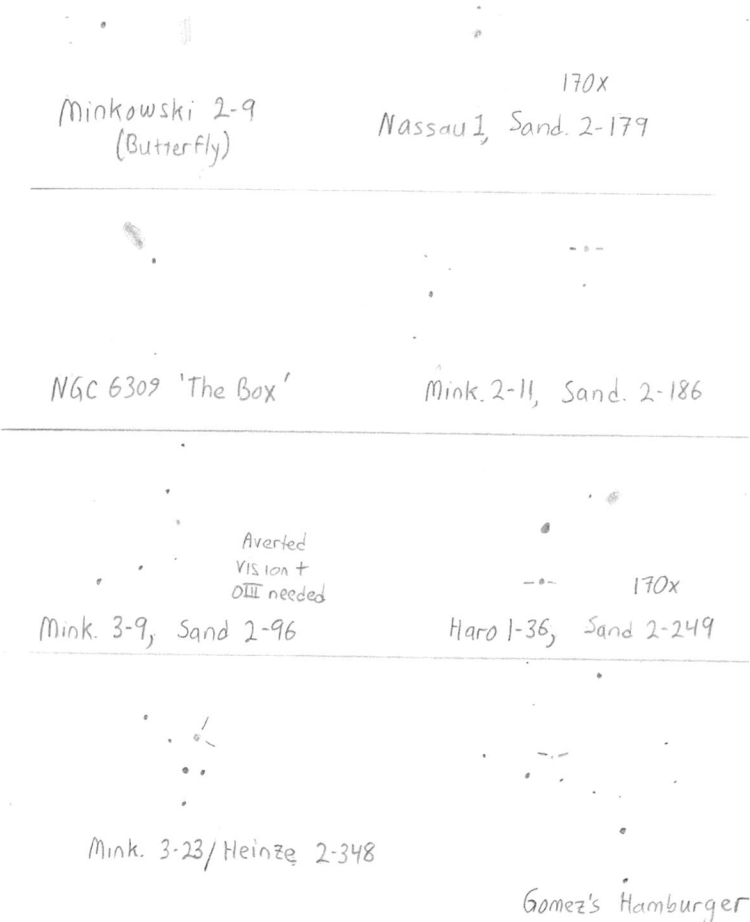

Fig. 9.1. Assorted sketches of faint objects (challenging planetary nebulae) observed at the 2003 Texas Star Party event. All eight sketches were made with a 10-in. (25.4-cm) f/6 reflector under (naked eye limiting) magnitude seven skies and magnifications of 85× to 170×. The identification is handwritten beneath each sketch (black on white).

the appendix of this book. One is encouraged to go to the included website and look for the information and lists if one chooses to take on this project. There may be readers who live in South Africa, South America, Australia or New Zealand who would like to pursue challenging observing projects and this information is for you. Also us northern hemisphere dwellers may, from time to time, take an astronomical vacation to the Southern Hemisphere, and for those of you who do so, this information is for you as well. If you succeed in at least partially completing the lists, you will have begun the fun and challenging practice of "intergalactic astronomy," that is, observing deep sky objects that are found in galaxies outside the Milky Way.

Of course, simply viewing other galaxies can qualify you for intergalactic observing, but there is something special and unique about observing an HII region (emission nebula) or a planetary nebula, or a star cluster that resides in another galaxy besides our own. Several of the observing lists in Chap. 10, as well as some of the lists you can find online and elsewhere, feature projects of hunting down objects in M51 or M101, or some other bright nearby galaxy. The galaxy itself may (or may not) be bright, which itself provides for some satisfying views through large telescopes under a dark, moonless sky; the objects it contains provides the stimulation of an observing challenge and the satisfaction of having seen rarely observed objects millions of light years away. One or two galaxies' worth of deep objects may keep one busy for most of a night.

Sky & Telescope magazine usually has an article on "Going Deep," which focuses on faint objects. Subscribing to this magazine brings a wealth of astronomical information and beautiful pictures to your mailbox (the traditional postal kind) each month. This article usually runs every other issue, but even the monthly article by Sue French usually includes a few faint and/or challenging objects. Finally, good planetarium software (such as those described in Appendix D of this book) comes with databases of tens to hundreds of thousands of deep sky objects to keep the enthusiastic deep sky astronomer busy for a few lifetimes.

Whatever you do, make it manageable and enjoy the process of looking at each object in your listing of objects that you choose to observe, enjoy the view, and record what you see. The next chapter gives several sample lists of hundreds of deep sky objects to get you started. Chapter 11 will go into more depth on what to record when you document your observations.

Chapter 10

Suggested Projects by Survey and Source

Introduction

In the last chapter was an overview of the variety of projects that are out there that one can use to get started, and keep going, as far as observing faint objects is concerned. In this chapter, we provide a group of projects containing 721 individual objects (including four groups of galaxies, eleven pairs of galaxies, and ten clusters of galaxies). These lists are provided in the form of 16 tables, each grouping containing information on various deep sky objects according to object type and where they are found (in the Milky Way, in the Andromeda Galaxy, in other galaxies, etc.). Each table represents a separate observing project, although some tables (such as Tables 10.1, 10.6, 10.14, and 10.15) are quite long, with many entries to keep one busy for a while. There are certainly enough objects listed here, collected from a variety of sources (and with references to more lists related to these and presented in Appendix E of this book), to keep the serious visual amateur hunting for faint objects for years to come.

When describing a process of observing deep sky objects, we proceed systematically through the list, starting with objects "close to home" and work our way outward to the more distant parts of the universe.

Inside the Milky Way Galaxy

Nebulae (13 objects, including 3 "bright" and 9 planetary nebulae)

Sharpless HII Regions (10 objects)

Ancient Giant Planetaries (11 objects, 2 overlap with "Nebulae" list)

PT (Palomar/Terzan) and Obscured Globular Clusters (30 objects)

The Local Group of Galaxies

Globular Clusters of the Andromeda Galaxy (M31, the brightest 75)

Extragalactic Globulars (8 objects, 3 overlap with M31 GC list; 1 is also found on the M33 Deep Space Objects list)

M33 HII Regions (bright or emission nebulae) and Star Clouds (31 objects)

HII Regions in External Galaxies (11 objects, 1 overlaps with M33 HII Region list)

Local Group Dwarf Galaxies (10 objects)

B. Cudnik, *Faint Objects and How to Observe Them*, Astronomers' Observing Guides,
DOI 10.1007/978-1-4419-6757-2_10, © Springer Science+Business Media New York 2013

Table 10.1. Challenge Sampler (Courtesy of Steve Gottlieb/"Adventures in Deep Space")

Name	Type[a]	RA	Dec	Size	VMag	Const.	Description
UGC 12914	GX-Scd:p	00 01.6	+23 29	2.3 × 1.3	12.4	Peg	The "Taffy Galaxies" are a close pair of edge-ons (along with UGC 12915)
NGC 3	GX-S0	00 07.3	+08 18	1.2 × 0.7	13.3	Psc	Brightest in a galaxy group with NGC 7838 6.3′ NW, NGC 7837 6.9′ NW, N7835 10′ NW, N7834 11′ WNW and NGC 4 5′ NNE
IC 10	GX-IBm	00 20.4	+59 18	6.3 × 5.1		Cas	Low surface brightnes dwarf in the local group. Situated in a very rich star field just 3.3° from the galactic plane
Abell 2	PN	00 45.6	+57 58	33″ × 29″	14.5	Cas	Abell planetary located just 30′ WNW of mag 3.4 Eta Cassiopeia. Use an OIII filter
IC 59	E + R	00 56.7	+61 04	10 × 5		Cas	Faint reflection nebula 20′ N of Gamma Cassiopeia with IC 63 20′ SE
EGB 1	PN	01 07.2	+73 33	300″ × 180″		Cas	Challenging planetary involved with a group of faint stars. Use an OIII filter
Simeis 22	PN	01 30.7	+58 23	10 × 3		Cas	Large, low surface brightness glow in a rich milky way field. Confirm with an OIII filter
IC 166	OC	01 52.4	+61 51	5	11.7	Cas	Difficult open cluster masquerading as a small milky way patch. Use high power to resolve
HFG 1	PN	03 03.8	+64 54	523″	12	Cas	Extremely low surface brightness, huge ancient planetary
Q0957 + 561	QSR	10 01.3	+55 54		16.5	UMa	Gravitationally lensed twin quasars (components 16.5–16.7 magnitude separated by 6″) just 15′ NNW from NGC 3079
UGC 5459	GX-SBc	10 08.2	+53 05	4.8 × 0.7	12.6	UMa	This extremely narrow edge-on has a very striking appearance as it hangs from a mag 8.5 star
HCG 56	GX-Chain	11 32.6	+52 57	1.1 × 0.3		UMa	Challenging interconnected galaxy chain located 7′ south of NGC 3718
NGC 3172	GX-SA0	11 47.3	+89 06	0.7 × 0.7	13.6	UMi	This is the closest NGC galaxy to North Celestial Pole and is known as "Polarissima Borealis"
UGC 7321	GX-Sd	12 17.6	+22 32	5.5 × 0.4	13.4	Com	This is one of the thinnest known galaxies (major/minor axis ratio)
Mrk 205	GX-Sy	12 21.7	+75 19	Stellar	14.5	Dra	This Seyfert galaxy appears as a mag 14.5–15 "star" less than 1′ south of N4319
IC 972	PN	14 04.4	−17 14	43″ × 40″	13.6	Vir	Relatively easy but little-known planetary in Virgo
UGC 9242	GX-Sd	14 25.3	+39 32	5.0 × 0.3	13.5	Boo	Super-thin ghostly streak may require high power
Palomar 5	GC	15 16.1	−00 07	6.9	11.8	Ser	Extremely low surface brightness globular located 30′ south of mag 5.6 4 Serpentis
AGC 2065	GXCL	15 22.7	+27 43	30	15.6	CrB	Distant and challenging Corona Borealis galaxy cluster; six brightest members nearly 16th magnitude
IC 1116	GXCL	15 21.9	+08 25	1.6 × 1.6	12.8	Ser	Brightest member of rich cluster Abell 2063 although it is 15′ SW of the main clump of galaxies. Up to two dozen small, faint galaxies visible in the region
IC 4553	GX-S	15 35.0	+23 30	1.5 × 1.2	13.2	Ser	Considered the prototype of a megamaser with 98 % of its emission in the infrared. This is an interacting double system with an extremely faint "knot" at the south end
NGC 6027	GX-S0p	15 59.2	+20 46	0.4 × 0.2	14.3	Ser	Seyfert's Sextet, an extremely compact group! Use high power to resolve three or more members
AGC 2151	GXCL	16 05.1	+17 45	40	13.8	Her	Rich Hercules Galaxy Cluster includes several trios and subgroups
Longmore 13	PN	16 09.8	−30 55	71″	15.5	Sco	Large, very low surface brightness glow

(continued)

Table 10.1. (continued)

Name	Type[a]	RA	Dec	Size	VMag	Const.	Description
NGC 6107	GX-E	16 17.3	+34 54	0.9 × 0.7	13.8	CrB	Brightest member of the N6107 cluster in Corona Borealis with eight NGC members
IC 4603	RN	16 25.4	−24 28	20 × 10		Oph	This large, circular glow surrounding a 4′ pair of mag 8/10 stars is part of the Rho Ophiuchi complex (2° N of Antares)
Abell 39	PN	16 27.5	+27 54	170″	12.9	Her	Relatively bright Abell planetary at least 2.5′ in diameter! Use an OIII filter
NGC 6166	GXCL	16 28.6	+39 31	40	13.9	Her	Brightest and largest in the rich, compact Abell Galaxy Cluster 2199. Surrounded by a swarm of faint, tiny galaxies
IC 4617	GX-Sb	16 42.1	+36 42	1.1 × 0.4		Her	Challenging galaxy just 14′ NNE of the core of M13 and 15′ SW of N6207!
MCG +09-27-094	GX-E1	16 49.2	+53 25	0.3 × 0.25	14.6	Dra	Brightest member of Shakhbazian 16, which consists of five galaxies in a 3.5′ chain oriented N-S. View severely hampered by a mag 9 star just 2.0′ NE!
MCG +14-08-017	GX-Chain	16 52.8	+81 38		14.9	UMi	Brightest in the UGC 10638 chain (Shakhbazian 166) located 30′ SE of mag 4.2 Epsilon UMi
M 2-9	PPN	17 05.6	−10 09	39″ × 15″	14.6	Oph	Minkowski's Butterfly nebula is a bi-polar proto-planetary, best viewed at high power. Look for two thin "jets" N-S
Draco Dwarf	GX-dE0	17 20.2	+57 55	35.5 × 24.5		Dra	The Draco Dwarf is a large, very low surface brightness member of the Local Group, ~15′ × 10′, elongated N-S
Barnard 72	DN	17 23.5	−23 38	30		Oph	The Snake Nebula is a challenging dark nebula in Ophiuchus
IC 1257	GC	17 27.1	−07 06	1	13.1	Oph	Small low surface brightness globular located 6′ W of a mag 11.5 star. Discovered to be a globular in 1996
Haute-Provence 1	GC	17 31.1	−29 59	2.9	12.5	Oph	This difficult obscured globular is just brighter than the background milky way glow
Abell 43	PN	17 53.6	+10 37	80″ × 74″	14.7	Oph	Moderately large Abell planetary at least 1′ diameter with a couple of superimposed stars. OIII filter helps
IC 4677	PN	17 58.3	+66 38			Dra	This is an ionized knot in the outer halo of the Cat's Eye PN, 1.7′ west of the center! Use over 200× and a UHC filter
Djorgovski 2	GC	18 01.8	−27 50	3.5	9.9	Sgr	Recently discovered globular just 21′ WNW of open cluster NGC 6520 and dark nebula B86!
Barnard 87	DN	18 04.3	−32 30	12		Sgr	The "Parrot Head" dark nebula contains a single mag 9.5 star. Look for a short thin extension (beak of the Parrot)
Sh 2-68	PN	18 25.0	+00 52	475″ × 330″	11.2	Ser	Huge low surface brightness glow, perhaps 5′−6′ in diameter with an OIII filter
Abell 46	PN	18 31.3	+26 56	63″ × 60″	13.8	Lyr	Challenging 1′ Abell planetary. Try for the 15th magnitude central star
Palomar 8	GC	18 41.5	−19 50	4.7	11.2	Sgr	One of the easier Palomar globulars, roughly 2′ in diameter. A large scope may partially resolve
IC 1296	GX-SBbc	18 53.3	+33 04	1.1 × 0.9	14	Lyr	This faint galaxy is situated just 4′ NW of M57 along the north side of a small rhombus of mag 13–14 stars with sides of 1.5′

(continued)

Table 10.1. (continued)

Name	Type[a]	RA	Dec	Size	VMag	Const.	Description
NGC 6717	GC	18 55.1	−22 42	3.9	9.2	Sgr	Very unusual small, faint glow just 2′ south of mag 5 Nu 2 Sagitarii!
Sh 2-71	PN	19 02.0	+02 09	124″ × 75″	12.3	Aql	Interesting, large PN with an OIII filter. Look for a faint extension on the south side, giving a N-S elongation and irregularies in the rim
NGC 6749	GC	19 05.3	+01 54	6.3	12.4	Aql	Very low surface brightness globular in a very dusty portion of the Aquila milky way. This cluster and NGC 6380 are probably the two most difficult NGC globulars
Abell 55	PN	19 10.5	−02 21	47″ × 32″	13.2	Aql	This fairly small Abell PN is slightly elongated SSW-NNE. Use an OIII filter
Terzan 7	GC	19 17.7	−34 40	2.6	12	Sgr	Low surface brightness glow with little, if any, central concentration (this globular may have captured from the Sagittarius dwarf Spheroidal galaxy)
Abell 61	PN	19 19.2	+46 15	201″	13.5	Cyg	This large Abell planetary spans 2.5′–3.0′ in diameter and is irregularly lit. Situated in a rich milky way field
Arp 2	GC	19 28.7	−30 21	2.5	12.3	Sgr	Extremely low surface brightness 2′ glow with a very small brighter core. Situated in a rich star field and difficult to pick out
Parsamyan 21	RN	19 29.0	+09 39		14.2	Aql	An unusual "cometary nebula" Parsamyan 21 is very small (~15″). Search 45′ NW of N6804. Look for a "tail" on the north side
Sh 2-91	SNR	19 35.6	+29 37	120 × 2		Cyg	Huge supernova wisp (discovered to be a SN remant in 1997), nearly ~13′ × 2′ oriented SW-NE passes through mag 8.5 HD 185735 (V2086 Cygni). Located 14′ south of mag 4.7 Phi Cygni
UGC 11466	GX-S	19 43.0	+45 18	2.0 × 1.2	11.7	Cyg	Relatively bright UGC galaxy just 23′ NW of Delta Cygni and 11° above the galactic plane
NGC 6822	GX-IBm	19 45.0	−14 48	15.5 × 13.5	8.8	Sgr	Use an OIII filter on Barnard's galaxy and search for small HII regions on the north side of the galaxy
Palomar 11	GC	19 45.2	−08 00	8	9.8	Aql	This Palomar globular appears as a diffuse, irregular glow 4′ SSE of a mag 9 star. Try to resolve at high power
Abell 65	PN	19 46.6	−23 09	134″ × 34″	13.8	Sgr	This relatively bright Abell planetary is noticeably elongated NW-SE and appears similar to a low surface brightness galaxy
Sh 2-84	EN	19 49.0	+18 23	15 × 3		Sge	The "Little California Nebula" appears a faint, 4′ shallow arc of nebulosity bracketed by two mag 8.5 stars. Located 25′ ESE of mag 3.7 Delta Sagittae
ESO 461-007	GX-S0	19 52.1	−30 49	1.2 × 0.7	13.3	Sgr	Brightest of four galaxies in HCG 86
Abell 70	PN	20 31.6	−07 05	45″ × 40″	14.7	Aql	Once you've tracked down this planetary, look for a slight brightening on the north side – that's an uncatalogued galaxy shining through the disc!
Gyulbudaghian's Nebula	RN	20 45.9	+67 58	20″ × 15″		Cep	Challenging reflection nebula surrounding the very young T Tauri star PV Cephei. Look for a small glow which fans out from the 15th magnitude star
ESO 597-036	GX-S0 pec	20 48.2	−19 51	1.6 × 0.3	14.3	Cap	Brightest of three or more galaxies in HCG 87 (Hickson compact group)
Abell 72	PN	20 50.1	+13 33	134″ × 121″	12.7	Del	Large, 2′ disc with a mag 8 star just off the WSW edge!

(continued)

Table 10.1. (continued)

Name	Type[a]	RA	Dec	Size	VMag	Const.	Description
Simeis 3-210	SNR	20 53.1	+ 29 39			Cyg	Thin filament at the extreme southern end of the Veil Nebula, passes through mag 6.4 star
CRL 2688	PPN	21 02.3	+ 36 42	24″ × 6″		Cyg	The bi-polar Egg Nebula appears as a small, faint double object at high power
vdB 142	RN/DN	21 36.7	+ 57 30	15		Cep	The "Elephant's Trunk" is an unusual cometary globule (associated with star formation) on the west side of the huge but faint HII complex, IC 1396. Look for a 15′ × 5′ lane using a UHC filter
IC 5146	EN	21 53.5	+ 47 16	12 × 12		Cyg	The Cocoon Nebula is a large milky glow surrounding two mag 9 stars. Try using an H-beta filter and look for the long dark lane which extends WNW
M 2-51	PN	22 16.1	+ 57 29	47″ × 38″	13.6	Cep	Relatively easy but little-known 30″ planetary responds well to a UHC or OIII filter
NGC 7320	GX-Sd	22 36.1	+ 33 57	2.2 × 1.1	12.6	Peg	Brightest in Stephan's Quintet. If the Quintet is easy, look for NGC 7320A 12′ SE, NGC 7320B 20′ E and NGC 7320C just 4′ E of NGC 7319!
Sh 2-155	EN	22 56.8	+ 62 37	50 × 30		Cep	Sh 2-155 is the "Cave Nebula" included by Patrick Moore in his "Caldwell Catalogue". Look for just a large, diffuse glow mostly surrounding a mag 8.5 star and a small knot 3′ ENE
NGC 7492	GC	23 08.4	− 15 37	4.2	11.5	Aqr	Low surface brightness NGC globular
Sh 2-157	EN	23 16.0	+ 60 28	60 × 50		Cas	This huge HII region appears as a faint, curving graceful arc, ~35′ × 8′, very elongated N-S and bowed out on the following side with an OIII filter. Extends north and south of open cluster Markarian 50 off the west side
Jones 1	PN	23 35.9	+ 30 28	314″	12.1	Peg	Huge annular planetary (nearly 5′ diameter) is dominated by two relatively narrow bright arcs in the rim along the NNW and SSE sides giving an unusual horseshoe or "C" shape
IC 5357	GX-S0pec	23 47.4	− 02 18	1.7 × 1.1	12.9	Psc	Brightest of four IC galaxies in HCG 97 = Shakhbazian 30
Abell 84	PN	23 47.8	+ 51 24	147″ × 114″	13	Cas	Look for a 2′ disc with an OIII filter with a star embedded on the east side

[a]Object type goes like so

"GX" stands for 'Galaxy', with the galaxy type following

"PN" is "Planetary Nebula"

"E + R" means "Emission and Reflection nebula"

"OC" is "Open Cluster"

"QSR" means "Quasar"

"GC" signifies "Globular Cluster"

"GXCL" stands for "Galaxy Cluster"

"RN" is "Reflection Nebula"

"PPN" is "Proto-Planetary Nebula"

"DN" stands for "Dark Nebula"

"SNR" is a "Super Nova Nebula"

"EN" is the abbreviation for "Emission Nebula"

Beyond the Local Group – Interacting Galaxies and Galaxy Groups
Compact Galaxy Groups Sampler (10 objects, 6 objects overlap with Hickson)
Arp Interacting Pairs Sampler (11 objects)
Abell Galaxy Clusters (10 Clusters)
Galaxies Galore! (87 objects)
Hickson Compact Groups (100 groups of galaxies containing 324 individual
objects)

Mix it Up/Going Deep
Challenge Sampler (75 objects)
Quasars, Blazars, and other Exotic Objects (18 objects)

The above includes 16 observing lists that are presented in this chapter. Some of these lists come with references to additional projects. These references are presented in Appendix E along with the websites on where to find each list and still more references for more lists, including exploration of the Large Magellanic Cloud for Southern Hemisphere observers.

We will start as an example with the wide variety of faint objects in the "Challenge Sampler" (Table 10.1). Included here are the object type, equatorial coordinates, the apparent size, the apparent visual magnitude, the constellation (three-letter abbreviation) it resides in, and comments from deep sky observer Steve Gottlieb. The types of objects run the gamut of what is out there and include galaxies of many types, planetary nebulae, emission or reflection nebulae, open clusters, globular clusters, quasars, galaxy clusters, proto-planetary nebulae, dark nebulae, and the supernovae remnants.

The objects are presented in order of right ascension, with a gap between 3 and 10 h R.A. The first batch (ranging from 0 to 3 h R.A.) are best observed during Northern Hemisphere autumn, with some of the more northerly objects visible year round. Objects residing between 10 and 15 h R.A. are best viewed in the evening sky during Northern Hemisphere spring; the ones situated from 16 to 21 h R.A. are at their best visibility throughout the summer in the evenings. The last of the objects, from 22 h and up, are seen during early autumn.

Now we move into the object-specific lists, beginning with the nebulae.

Nebulae

This list of challenging planetary nebulae (Table 10.2, with some "bright" nebulae thrown in for good measure) was cobbled together from a variety of sources. Most can be seen with 10- to 12-in. telescopes and a nebula filter, but some are a bit more challenging. "CSMag" is the central star magnitude. The nebula are presented in order of right ascension so that one can get an idea as to which season (or time of night) the object is best observed by noting its host constellation (it's a simple Internet search to find out when the constellation is high in the sky during the evening at your location, or you can use your favorite planetarium program or sky software to accomplish this purpose). Most of these objects are best seen during the summertime months from the Northern Hemisphere during the evening hours.

Table 10.2. Nebulae, sources referenced within

Name	Type	RA	Dec.	Const.	VMag.	CS Mag.[a]	Size	Source/comments
Cederblad 214	BN	00 02 00	+67 02 00	Cep	–	–	60′	
Abell 4	PN	02 45 23.7	+42 33 05	Per	14.8	19.9	22″	
Purgathofer Weinberger 1	PN	6 19 34.3	+55 36 43	Lynx	11.2	15.4	20′	
IC 972	PN	14 04 25.9	−17 13 41	Vir	13.6	17.9	54″	
Abell 39	PN	16 27 33.7	+27 54 33.4	Her	13.7	15.5	155″ × 154″	
Minkowski 2-9	PN	17 05 37.9	−10 08 34.6	Oph	14.7	15.7	39″ × 15″	
Sharpless 2-68	BN	18 24 59.8	+00 51 58	Ser	13.1	16.5	430″ × 330″	
Minkowski 1-59	PN	18 43 20.2	−9 4 49	Sct	12.4	–	4.8″ × 4.3″	
NGC 6742	PN	18 59 19.9	+48 27 56	Dra	13.4	20	31″ × 30″	
Pease 1	PN	21 30 02	+12 10 12	Peg	15.5	–	3″	Located within M15
NGC 7094	PN	21 36 52.9	+12 47 16	Peg	13.4	–	94″	
Minkowski 1-79	PN	21 37 00.6	+48 56 12.1	Cyg	13.2		38″ × 27″	
Sharpless 2-140	BN	22 15 54	+62 47 00	Cep	–	–	>350′	

[a]"CS Mag." is central star magnitude

Table 10.3. Sharpless HII regions (never a dull moment with these!) (Courtesy of Steve Gottlieb/"Off the Deep End – 10 Challenging Observing Projects")

Name	R.A.	Dec	Size	Const.	Description
Sh 2-11	17 24.7	−34 12	50′ × 40′	Sco	The brightest section of NGC 6357 is distinctive, elongated strip with a close double at the S edge. Faint open cluster Pismis 24 to south
Sh 2-54	18 17.9	−11 44	6′	Ser	Obscure but relatively easy HII nebula located 30′ north of NGC 6604 and Sh 2–54 (large, faint HII region encasing the cluster)
Sh 2-84	19 49.0	+18 23	15′ × 3′	Sge	The "Little California Nebula" appears a faint, 4′ shallow arc of nebulosity bracketed by two mag 8.5 stars. Located 25′ ESE of mag 3.7 Delta Sagittae
Sh 2-105	20 12.1	+38 21	20′ × 10′	Cyg	Using 100× and OIII filter in an 18″, the Crescent Nebula (NGC 6888) is a huge 18′ × 11′ oval, with knots and tendrils, in a rich Cygnus star field
Sh 2-108	20 22.2	+40 15		Cyg	IC 1318 is an extremely large and faint HII complex around Gamma Cyg. Use very low × and H-beta or UHC filter
Sh 2-112	20 33.9	+45 39	9′ × 7′	Cyg	Very irregular, relatively easy HII region, 10′ diameter, surrounding mag 9 star
Sh 2-125	21 53.5	+47 16	12′ × 12′	Cyg	The Cocoon Nebula (IC 5146) is a large milky glow surrounding two mag 9 stars. Use an H-beta filter. Look for a long dark lane extending WNW
Sh 2-132	22 19.1	+56 05	30′ × 20′	Cep	With an OIII filter this huge HII region appears roughly 20′ × 15′, elongated E-W
Sh 2-157	23 16.0	+60 28	60′ × 50′	Cas	Huge HII region is a faint curving arc, ~35′ × 8′, very elongated N-S and bowed out on the following side. Extends N and S of cluster Mrk 50 off the W side
Sh 2-162	23 20.7	+61 12	15′ × 8′	Cas	The Bubble Nebula (NGC 7635) is a low surface brightness HII region that curves mostly north of a mag 8.5 illuminating star. Located 35′ SW M52

This is by no means a complete listing. Some of the sources mentioned in the previous chapter will provide additional challenges to satisfy any diehard deep space faint object aficionado! The next two lists of such challenging objects are grouped by nebula type: emission nebulae (or HII regions, Table 10.3) cataloged by Sharpless, and planetary nebulae, the large, ancient giant type that have low surface brightness and are generally hard to see (Table 10.4).

Table 10.4. Ancient giant planetaries (Courtesy of Steve Gottlieb/"Off the Deep End — 10 Challenging Observing Projects.")

Name	RA	Dec	Size	VMag	Const.	Description
EGB 1	01 07.2	+73 33	300″ × 180″		Cas	Challenging planetary involved with a group of faint stars. Use an OIII filter and low power!
HFG 1	03 03.8	+64 54	523″	12	Cas	Extremely low surface brightness, huge ancient planetary, perhaps 8′ in diameter
HDW 2	03 11.0	+62 48	340″		Cas	Huge, very low surface brightness glow surrounding three mag 11 stars, ~4′ in diameter. Use OIII filter and low power
Abell 39	16 27.5	+27 54	170″	12.9	Her	Relatively bright Abell planetary at least 2.5′ in diameter! Use an OIII filter
Sh 2-68	18 25.0	+00 52	475″ × 330″	11.2	Ser	Huge low surface brightness glow, perhaps 5′–6′ in diameter with an OIII filter
Sh 2-71	19 02.0	+02 09	124″ × 75″	12.3	Aql	Large, interesting PN with an OIII filter. Look for faint extension on the S side, giving a N-S elongation and irregularities in the rim
Abell 61	19 19.2	+46 15	201″	13.5	Cyg	Large Abell planetary spans 2.5′–3.0′ in diameter and irregularly lit. Situated in a rich Milky Way field
Abell 65	19 46.6	−23 09	134″ × 34″	13.8	Sgr	Relatively bright Abell planetary, noticeably elongated NW-SE, appears similar to a low surface brightness galaxy
Abell 72	20 50.1	+13 33	134″ × 121″	12.7	Del	Large, 2′ disc with a mag 8 star just off the WSW edge!
Jones 1	23 35.9	+30 28	314″	12.1	Peg	Huge annular planetary with two brighter arcs in the rim along the NNW and SSE sides giving a horseshoe or "C" shape
Abell 84	23 47.8	+51 24	147″ × 114″	13	Cas	Look for a 2′ disc with an OIII filter with a star embedded on the east side

Just like the previous list of nebulae, the lists of Tables 10.3 and 10.4 contain objects best viewed during the summer evenings in the Northern Hemisphere when many observers do most of their observing. All four lists of objects best placed in the evening during the summer can also be seen predawn during the winter months, when the skies tend to be more transparent but the air is much, much chillier.

Next our gaze turns toward the great balls of stars that populate our galaxy's halo. These are the more obscure versions of the familiar objects, such as M13 and M22, which can be quite difficult to see.

Globular Star Clusters of the Milky Way

The Milky Way Galaxy hosts 160 globular clusters, many of which are easily seen through backyard 'scopes (Table 10.5). Many recognize the likes of M13, M22, M5, and so forth, but how many are familiar with Terzan 3 or Palomar 7? How about Djorgovski 2? Take a trip off the beaten galactic path to enjoy some of these less-oft visited friends. Each provides its unique challenge. Try to spot the component stars of each.

Table 10.5. Globular star clusters, including Palomar/obscured globulars, some from the "Off the Deep End — 10 Challenging Observing Projects" observing lists of Steve Gottlieb; others from sources referenced within

NAME	R.A. (2000.0)	Dec.	Const.	Mag., overall	Mag., brightest	Size	Source/comments
Palomar 1	03 33 23.0	+79 34 50	Cep	13.6	16.3	2.8'	[1]
Palomar 2	04 46 05.8	+31 22 55	Aur	13	18.8	2.2	[1]
Palomar 3	10 05 31.4	+00 04 17	Sex	13.9	18	1.6	[1]
Palomar 4	11 29 16.8	+28 58 25	UMa	14.2	18	1.3	[1]
Palomar 5	15 16 05.3	−00 06 41	Ser	11.8	15.5	6.9	[1] Extremely low surface brightness globular located 30' south of mag 5.6 4 Serpentis
Palomar 6	17 43 42.2	−26 13 21	Oph	11.6	na	1.2	[1]
Palomar 7	18 10 44.2	−07 12 27	Ser	10.3	15.7	7	[1] IC 1276 is an irregular 3' glow with a mag 13.5 star at the W edge. Has a mottled, patchy appearance
Palomar 8	18 41 29.9	−19 49 33	Sgr	11.2	15.4	4.7	[1] One of the easier Palomar globulars, roughly 2' in diameter. A large scope may partially resolve
Palomar 9	18 55 06.0	−22 42 06	Sgr	9.2	14	3.9	[1] NGC 6717 is a very unusual small, faint glow just 2' south of mag 5 Nu 2 Sagitarii!
Palomar 10	19 18 02.1	+18 34 18	Sge	13.2	18	4	[1]
Palomar 11	19 45 14.4	−08 00 26	Aql	9.8	na	8.0	[1] This Palomar globular appears as a diffuse, irregular glow 4' SSE of a mag 9 star. Try to resolve at high power
Palomar 12	21 46 38.8	−21 15 03	Cap	11.7	14.6	2.9	[1] Appears as an irregular 3' glow close NNW of a trio of mag 12 stars. May have been captured from the Sagittarius Dwarf Spheroidal galaxy
Palomar 13	23 06 44.4	+12 46 19	Peg	13.8	17	0.7	[1]
Palomar 14	16 11 04.9	+14 57 29	Her	14.7	17.6	2.5	[1]
Palomar 15	16 59 51	−00 32 31	Oph	14.2	17.1	3	[1]
Terzan 1	17 35 47.2	−30 28 54	Sgr	15.9	–	2.4	[2]
Terzan 2	17 27 33.1	−30 48 8	Sgr	14.3	–	1.5	[2]
Terzan 3	16 28 40.1	−35 21 13	Sco	12	15	3.3	[2]
Terzan 4	17 30 39	−31 35 44	Sco	16	–	0.7	[2]
Terzan 5 = Terzan 11	17 48 4.9	−24 46 45	Sgr	13.9	20.5	2.1	[2]
Terzan 6	17 50 46.4	−31 16 31	Sgr	13.9	20.5	1.2	[2]
Terzan 7	19 17 43.7	−34 39 27	Sgr	12	15	2.6	[2] Low surface brightness glow with no central concentration. This gc may have been a member of the Sagittarius Dwarf Spheroidal galaxy
Terzan 8	19 41 45	−34 0 1	Sgr	12.4	15	5.0	[2] Very faint, round glow, just 1.0'–1.5' diameter. Lies close to the recently discovered (1994) Sagittarius Dwarf Spheroidal galaxy
Terzan 9	18 1 38.8	−26 50 23	Sgr	16		1.5	[2]
Terzan 10	18 2 57.4	−26 4 0	Sgr	14.9	19.7	0.3	[2]
Terzan 12	18 12 15.8	−22 44 31	Sgr	15.6	18.5	1.5	[2]
Tonantzintla 1	17 34.5	−39 04	Sco				[3]
Tonantzintla 2	17 36.2	−38 33	Sco				[3]
Djorgovski 2	18 01.8	−27 50	Sgr	9.9	–	3.5	[3] Recently discovered globular just 21' WNW of open cluster NGC 6520 and dark nebula B86!
Arp 2	19 28.7	−30 21	Sgr	12.3		2.5	[3] Extremely low surf brightness glow, 2' diameter in a rich star field. Very difficult to pick out

Sources:

[1] Palomars from http://www.astronomy-mall.com/Adventures.In.Deep.Space/palglob.htm

[2] Terzans from http://natkobajic.netfirms.com/listTER7AN html

[3] Sky & Telescope, June 2002, p. 108

Object descriptions of selected clusters by Steve Gottlieb

The objects are listed in order of name, rather than right ascension, so there is some jumping back and forth in right ascension. But most, if not all, of the objects on the list are best seen during the late spring through early autumn timeframe because most globulars tend to congregate in the hemisphere of the celestial sphere that is on the galactic center. One of the few exceptions to this is M79 in Orion, a brighter winter globular cluster. Keeping with the theme of "faint objects," most of these objects are quite faint and obscure, and a few of them are far enough south to be difficult objects from the Northern Hemisphere, due to their low elevations (at best) in the southern sky when they are at their highest.

The list gives 30 (out of 160) of the most challenging planetaries to reside in the Milky Way. Most of the remaining 130 or so objects are more easily observed (a good number are only seen from the Southern Hemisphere, and Palomar 1 is an exclusively Northern Hemisphere globular).

Deep Space Objects in Other Galaxies

We have seen our share of deep space objects in our own galaxy, objects such as emission and planetary nebulae, open and globular star clusters, supernovae remnants and proto-planetary nebulae. Now we turn our gaze to the nearby galaxies, starting with M31, the great Andromeda Galaxy. Table 9.6 is a list of the 75 brightest globular clusters in M31. Go to the web site http://www.astronomy-mall.com/Adventures.In.Deep.Space/gcm31.htm for a description of the globulars in M31 as well as narratives about some of the open clusters and one stellar association/star cloud in the Andromeda Galaxy. You can also find charts of sections of M31 to help you confirm the identities of these objects on this website http://ned.ipac.caltech.edu/level5/ANDROMEDA_Atlas/frames.html.

Table 10.7 samples globulars from other galaxies, not just M31. Table 10.8 turns one's focus to the Triangulum Pinwheel Galaxy, M33, and some of the brightest HII regions and star clouds there. This comes with a finder chart (Fig. 10.1) that will make it easier to find and identify these objects. Finally, Table 10.9 highlights some HII regions that can be spotted in other galaxies. The galaxies M31 and M33 can best be seen during the autumn months, so much of what Tables 10.6, 10.7, 10.8 10.9 present are best seen during the autumn. The objects in Table 10.9 are visible during various times of the year: the first three objects are best seen during the autumn, and the next six can be spotted from the latter part of the winter into much of spring; the final two are visible during the summer months.

M33 HII Regions, star clouds, and other objects

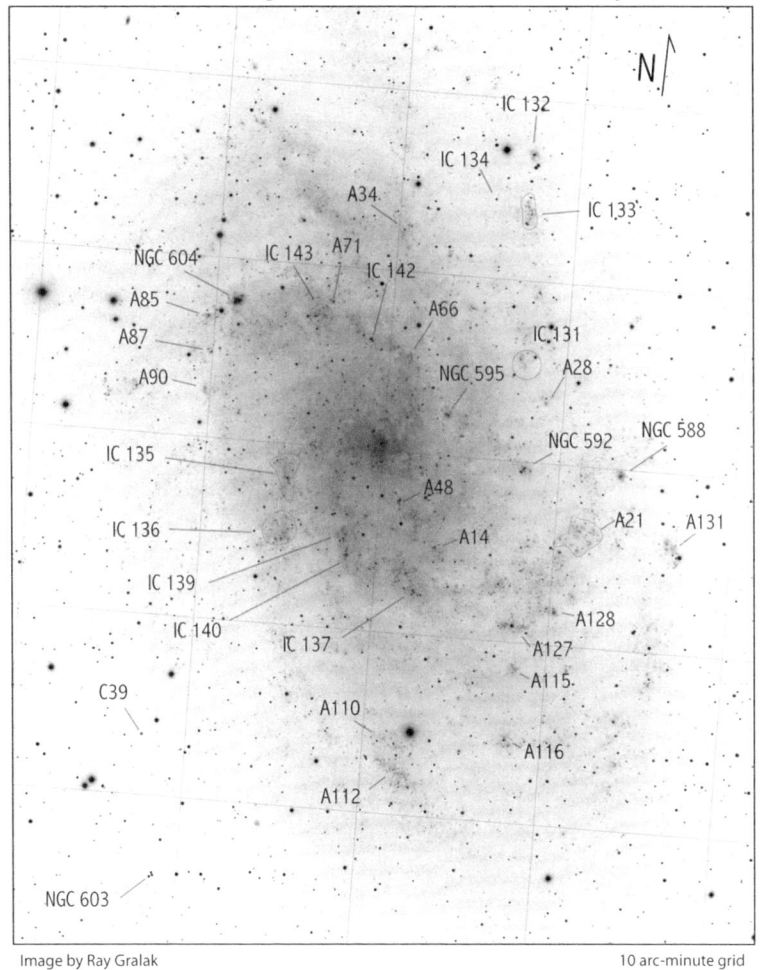

Image by Ray Gralak · 10 arc-minute grid

Fig. 10.1. M33 HII regions, star clouds and other objects (Courtesy of Ray Gralak and Steve Gottlieb).

Table 10.6. M31 globular clusters (the brightest 75) (Courtesy of Steve Gottlieb/"Adventures in Deep Space")

Name	Other	RA	Dec	VMag	Size
G001	Mayall II	00 32 47	+39 34 40	13.8	36″
G280	Bol 225	00 44 30	+41 21 37	14.2	2.7″
G078	Bol 023	00 41 01	+41 13 45	14.2	3.2″
G076	Bol 338	00 40 59	+40 35 47	14.2	3.6″
G185	Bol 127	00 42 44	+41 14 42	14.5	
G213	Bol 158	00 43 14	+41 07 21	14.7	2.5″
Bol 124	VDB 2	00 42 41	+41 15 24	14.8	
G272	Bol 218	00 44 14	+41 19 20	14.8	3.4″
G205	Bol 151	00 43 10	+41 21 33	14.8	2.9″
G073	Bol 020	00 40 55	+41 41 25	14.9	
G072	Bol 019	00 40 53	+41 18 53	14.9	2.2″
G119	Bol 058	00 41 53	+40 47 09	15.0	2.7″
G229	Bol 178	00 43 31	+41 21 16	15.0	3.4″
G217	Bol 163	00 43 18	+41 27 46	15.0	2.6″
G257	Bol 206	00 43 59	+41 30 18	15.1	3.2″
G064	Bol 012	00 40 32	+41 21 44	15.1	2.3″
G148	Bol 086	00 42 19	+41 14 01	15.2	2.9″
Bol D195	VDB 0	00 40 29	+40 36 15	15.2	
Bol 472		00 43 48	+41 26 53	15.2	
G351	Bol 405	00 49 40	+41 35 29	15.2	
Bol D091		00 43 01	+41 30 17	15.2	
G302	Bol 240	00 45 25	+41 06 24	15.2	2.5″
G219	Mayall IV	00 43 18	+39 49 14	15.2	
G165	Bol 103	00 42 30	+41 17 57	15.2	2.9″
G172	Bol 110	00 42 33	+41 03 28	15.3	2.4″
G222	Bol 171	00 43 26	+41 15 37	15.3	3.2″
G244	Bol 193	00 43 46	+41 36 58	15.3	2.6″
G318	Bol 383	00 46 12	+41 19 43	15.3	
G230	Bol 179	00 43 31	+41 18 14	15.4	2.9″
G150	Bol 088	00 42 21	+41 32 14	15.4	2.7″
G233	Bol 182	00 43 37	+41 08 12	15.4	2.6″
G189	Bol 131	00 42 51	+41 17 08	15.4	
G033	Bol 311	00 39 34	+40 31 14	15.4	2.3″
G279	Bol 224	00 44 27	+41 28 51	15.4	4.9″
G256	Bol 205	00 43 58	+41 24 38	15.5	3.1″
G096	Bol 034	00 41 28	+40 53 49	15.5	2.7″
G226	Bol 174	00 43 30	+41 38 57	15.5	3.8″
G263	Bol 212	00 44 03	+41 04 57	15.5	2.5″
G058	Bol 006	00 40 26	+41 27 26	15.5	2.3″
G035	Bol 312	00 39 40	+40 57 02	15.5	2.9″
G144	Bol 082	00 42 16	+41 01 14	15.5	2.8″
G235	Bol 185	00 43 37	+41 14 43	15.5	2.5″
G322	Bol 386	00 46 27	+42 01 53	15.5	2.3″
G156	Bol 094	00 42 25	+40 57 17	15.6	2.5″
G087	Bol 027	00 41 15	+40 55 51	15.6	2.9″
G305	Bol 373	00 45 42	+41 45 35	15.6	2.2″
G124	Bol 063	00 42 01	+41 29 10	15.7	2.3″
G286	Bol 232	00 44 40	+41 15 01	15.7	2.5″
G052	Bol 005	00 40 20	+40 43 58	15.7	2.3″
G254	Bol 204	00 43 57	+41 22 03	15.8	

(continued)

Table 10.6. (continued)

Name	Other	RA	Dec	VMag	Size
G319	Bol 384	00 46 22	+40 17 00	15.8	
G315	Bol 381	00 46 06	+41 21 00	15.8	
G287	Bol 233	00 44 42	+41 43 56	15.8	2.2″
G108	Bol 045	00 41 43	+41 34 21	15.8	2.5″
G002	Mayall III	00 33 34	+39 31 19	15.8	
G157	Bol 095	00 42 26	+41 05 36	15.8	
Bol 144		00 43 00	+41 16 06	15.9	
G250	Bol 201	00 43 53	+41 09 58	15.9	3.4″
G169	Bol 107	00 42 31	+41 19 39	15.9	2.4″
G327	Mayall VI	00 46 50	+42 44 45	15.9	2.7″
G070	Bol 017	00 40 49	+41 12 07	15.9	2.3″
G127	Bol 344	00 42 03	+41 52 03	15.9	
G234	Bol 183	00 43 37	+41 02 02	15.9	2.7″
G101	Bol 039	00 41 38	+41 20 50	16.0	2.5″
G134	Bol 073	00 42 07	+40 59 21	16.0	2.7″
G291	Bol 366	00 44 47	+42 03 50	16.0	3.4″
G198	Bol 143	00 43 00	+41 19 19	16.0	2.4″
G231	Bol 180	00 43 32	+41 07 46	16.0	2.5″
G168	Bol 106	00 42 31	+41 12 18	16.0	2.5″
G177	Bol 115	00 42 34	+41 14 02	16.0	2.5″
G192	Bol 135	00 42 52	+41 31 08	16.0	2.3″
G200	Bol 148	00 43 04	+41 18 05	16.1	2.3″
G283	Bol 230	00 44 35	+40 57 13	16.1	2.5″
G352	Bol 407	00 50 10	+41 41 01	16.1	
G114	Bol 051	00 41 47	+41 25 19	16.1	2.6″

Table 10.7. Extragalactic globulars (Courtesy of Steve Gottlieb/"Off the Deep End – 10 Challenging Observing Projects")

Name	RA	Dec	Size	VMag	Con	Description (by Steve Gottlieb)
WLM-1	00 01.8	−15 28			Cet	The brightest gc in the Wolf-Lundmark-Melotte system appears as just a mag 16 "star" on the W side of the galaxy
M31-G1	00 32.8	+39 35	10″	13.7	And	Brightest globular in M31 located 2.5° SW of the center. Appears as an out of focus "star" forming a small triangle with two mag 12/13 stars
M31-G78	00 41.0	+41 14	3.2″	14.2	And	Appears as a "soft" mag 14.5 star just 30′ W of the core of M31. This gc is the 2nd or 3rd brightest in M31
M31-G76	00 41.0	+40 36	3.6″	14.3	And	Easily visible as a mag 14.2 stellar object in a small "W" asterism and near the SW tip of M31. Similar brightness to G78
M33-C39	01 34.8	+30 22		15.9	Tri	Brightest globular cluster in M33 though only appears as an extremely faint mag 16 "star". Needs a finder chart to pin down location
NGC 1049	02 39.8	−34 15	1.3	12.6	For	This small 12th magnitude gc is the brightest of 5 in the Fornax Dwarf
Fornax-4	02 40.1	−34 32	0.8	13.6	For	Third brightest globular in the Fornax Dwarf, roughly mag 13.5 and very small
Fornax-5	02 42.3	−34 06	1.0	13.4	For	Second in brightness gc within the Fornax Dwarf and similar to Fornax-4

Table 10.8. M33 HII regions and star clouds (Courtesy of Steve Gottlieb/"Adventures in Deep Space")

Name	Other name	RA (J2000)	Dec (J2000)	VMag	Size
A131	BCLMP 274	01 32 30.8	+30 35 07	14.7	
NGC 588	A27 = BCLMP 280	01 32 45.6	+30 38 50		0.5′
A21	BCLMP 231/232/234/235/236/237	01 32 56	+30 35.6		
A128	BCLMP 218	01 33 01	+30 30.9		
IC 131	A29 = BCLMP 290	01 33 14.6	+30 44 56		
A115	BCLMP 220/221	01 33 11.5	+30 27 31		
A116	BCLMP 255/256/257/258	01 33 11.7	+30 23.4		
A127	BCLMP 216	01 33 11.9	+30 29 54		
NGC 592	A59 = BCLMP 277	01 33 12.5	+30 38 50		0.8′ × 0.7′
IC 132	BCLMP 638	01 33 15.8	+30 56 45		0.8′ × 0.6′
IC 133	A137 = BCLMP 624	01 33 15.8	+30 53 05	14.3	
NGC 595	A62 = BCLMP 49	01 33 33.5	+30 41 31		
IC 137	A12 = BCLMP 21/200/201/202/203/204/ 205/207/208	01 33 39.1	+30 31 20		
A14	BCLMP 209	01 33 34.6	+30 33 56		
A112	BCLMP 24/248/249/250/251	01 33 39	+30 21.0		
A48	BCLMP 25	01 33 45.0	+30 36 28		
A66	BCLMP 61/62	01 33 45	+30 44.7		
U49	Kron-Mayall "a"	01 33 45.0	+30 47 47	16.2	
IC 142	A67 = BCLMP 301	01 33 55.6	+30 45 26		0.5′
IC 139	A4 = BCLMP 6/7/11	01 33 59.2	+30 34 03		
IC 140	A5	01 33 58.1	+30 33 02		
A71	BCLMP 302	01 34 06.9	+30 47 26		
IC 143	A75 = BCLMP 688/689	01 34 11.2	+30 46 38		
IC 135	A100 = BCLMP 28/88	01 34 15.8	+30 37 11		
IC 136	A101 = BCLMP 88	01 34 17	+30 34.0		0.6′
NGC 604	A84 = BCLMP 680	01 34 31.9	+30 47 13		
A87	BCLMP 748/749/750	01 34 38.9	+30 44 00		
NGC 603	(Triple Star)	01 34 44.0	+30 13 58		
A90	BCLMP 740	01 34 39.6	+30 41 52	15.6	
A85	BCLMP 677/678/681	01 34 40.4	+30 46 01	15.8	
C39	Kron-Mayall "c"	01 34 49.2	+30 21 51	15.9	

Notes:
A# from Humphreys and Sandage, ApJS, 44, pp. 319–381 (1980)
BCLMP# from Boulesteix, Courtex, Laval, Monnet and Petit, A&A. 37, pp. 33–48 (1974)
U49 and C39 are globular clusters in M33
NGC 603 is a triple star found by Lord Rosse

The Local Group of Galaxies

We then move into intergalactic space with this next list, a challenging batch of faint, low surface brightness local group dwarf galaxies (Table 10.10). These are among the closest galaxies to us in the universe but are also among the most challenging due to their low surface brightness. In some cases, any scattered light from the Moon or from distant artificial light sources, as well as less than perfect transparency, are enough to render a barely visible object totally invisible.

Table 10.9. HII regions in external galaxies (Courtesy of Steve Gottlieb/"Off the Deep End — 10 Challenging Observing Projects")

Name	RA	Dec	Size	Con	Remarks
NGC 55	00 14.9	−39 12		Scl	Near the "core" of NGC 55 are two small HII knots. Additional fainter knots are along the fainter eastern end
NGC 253	00 47.6	−25 17		Scl	Remarkable dust structure and mottling visible particularly on the SW extension where three faint HII knots are visible near an embedded star
NGC 604	01 34.5	+30 47	1.9′ × 1.2′	Tri	Bright HII region 12′ NE core of M33. Also look for NGC 588, 592, 595, IC 131, 133, 135, 136, 137, 139, 140, 142, 143
NGC 2366	07 28.9	+69 13	0.4′	Cam	At the southwest end of the low surface brightness galaxy, NGC 2366, is a very bright HII knot, ~15″ in size which responds to a UHC filter!
NGC 2404	07 37.1	+65 42		Cam	Brightest of several HII regions in NGC 2403. Located close E of the core along the northern spiral arm, 1.5′ from the center
M108	11 11.5	+55 40		Cam	A few brighter patches or knots are visible along the major axis of M108 with a prominent HII knot along the west side
NGC 4401	12 25.9	+33 32		CVn	Brightest of several HII knots in NGC 4395, located ~2′ SE of the ill-defined core. Appears as an irregular 25″ knot
NGC 4449	12 28.2	+44 06		CVn	Several giant HII regions are visible, the brightest is a well-defined patch on the N edge of the galaxy (Hodge-Kennicutt 15), 1.5′ from the center
NGC 4861	12 59.0	+34 51	0.35′ × 0.30′	CVn	High surface brightness 15″ knot at the SSW end of NGC 4861. Appears more prominent than the low surface brightness galaxy!
NGC 5471	14 04.5	+54 24	0.9′ × 0.7′	UMa	Highest surface brightness of several HII regions in M101 on the extreme NE end. Others include NGC 5447, 5449, 5451, 5453, 5455, 5458, 5461, 5462
IC 1308	19 45.1	−14 43	1.0′ × 0.9′	Sgr	Following of two HII regions at the north end of Barnard's galaxy. A mag 12 star lies 2′ SE. Mild contrast gain with an OIII filter

It's not surprising that these objects are best seen in the fall (mostly) and the spring (a couple), when the parts of the sky away from the band of the Milky Way are best seen. The Milky Way, with its obscuring dust, gas, and stars, makes it difficult to impossible to see any other galaxies beyond, with a few exceptions. Two of these exceptions are listed in Table 10.10: SagDIG and NGC 6822, each of which are found in rich Milky Way fields of view, which makes them even more challenging than what they otherwise would be.

Interacting Galaxies and Groups of Galaxies

The focus for the next set of tables is galaxies, as they appear by themselves or in pairs, or in compact groups or clusters of galaxies. All of the tables (Tables 10.10, 10.11, 10.12, 10.13 and 10.15), except 10.14, are courtesy of Steve Gottlieb; Table 10.14 is a list of galaxies and small groups of galaxies that have been assembled from a variety of sources, primarily from back issues of *Sky & Telescope* magazine. The references are provided within the table itself.

Table 10.10. Local Group dwarf galaxies (Courtesy of Steve Gottlieb/"Off the Deep End – 10 Challenging Observing Projects")

Name	RA	Dec	Size	VMag	Con	Notes
WLM System	00 01.9	−15 27	11.5′ × 4.0′	10.6	Cet	The WLM system is a difficult, very low contrast galaxy in Cetus spread out over 10′ × 5′, elongated N-S without a central concentration
IC 10	00 20.4	+59 18	6.3′ × 5.1′	11.0	Cas	Appears as a large, low surface brightness hazy region with no concentration and a mag 13 star superimposed. Only 4° from the galactic plane
NGC 147	00 33.2	+48 30	13.2′ × 7.8′	9.5	Cas	Satellite system of M31 with a very low surface brightness
NGC 185	00 39.0	+48 20	11.7′ × 10.0′	9.2	Cas	Relatively bright and large satellite system of M31. The brightest globular is only mag 16.5 and located 8′ N of center
IC 1613	01 04.8	+02 07	16.2′ × 14.5′	9.2	Cet	The Cetus system is a local group member at 2.4 million l.y. and appears as an irregular hazy region with a brighter section to the Northeast
Ursa Minor Dwarf	15 08.8	+67 11	41′ × 26′	10.9	UMi	Challenging dwarf elliptical has an extremely low surface brightness. Look for subtle brightening in a low power field
Draco Dwarf	17 20.2	+57 55	35.5′ × 24.5′	10	Dra	The Draco Dwarf is a large, very low surface brightness member of the Local Group, ~15′ × 10′, with several stars superimposed
SagDIG	19 30.0	−17 41	3.2′ × 1.5′	15	Sgr	The Sagittarius Dwarf Irregular is a difficult 3′ roundish glow nestled in a rich Milky Way field. Very difficult to identify
NGC 6822	19 45.0	−14 48	15.5′ × 13.5′	8.8	Sgr	Barnard's Galaxy has a low, but irregular surf br and appears ~14′ in length. Use an OIII filter to search for two small HII regions on the North end
Pegasus Dwarf	23 28.6	+14 45	5.0′ × 2.7′	12.6	Peg	This challenging dwarf is just a very low surface brightness hazy region with no concentration, roughly 4′ × 2′

Note: Observation Notes by Steve Gottlieb

Keep in mind as you view these special groupings, in many cases, two galaxies are in the process of "crashing into" each other. This is a dynamic event that is happening before our eyes, but because of the immense sizes of the galaxies, to our senses the events are taking place in extremely slow motion. In fact when we look at various interacting pairs of galaxies we are literally seeing snapshots of various stages of collision between pairs of galaxies. To get a "more complete" picture of a galaxy collision, one would need to either observe many galaxies in various stages of collisions (which would be akin to viewing photos of various automobile collisions taking place in a variety of settings with a variety of initial conditions and a variety of vehicle types) or one would need to simulate the collision on a supercomputer. At least the observer would actually get the feel for the dynamic nature of the interaction.

Nonetheless consider the action that is taking place at the sight of the collision as you ponder the two touching gray blurs in your eyepiece: lots of stars passing by each other and gas clouds colliding and compressing, setting the stage for explosive star birth at some point in the "near" future in one or both colliding galaxies.

Table 10.11. Compact galaxy groups (Courtesy of Steve Gottlieb/"Off the Deep End – 10 Challenging Observing Projects")

Name	RA	Dec	VMag	Con	Comments
Hickson 61	12 12.3	+29 11	12.2	Com	One of my favorite Hicksons – known as "The Box". Four galaxies, NGC 4169, 4173, 4174, 4175, form a rectangle with sides 2.5′ by 1.5′
Hickson 68	13 53.4	+40 17	11	CVn	One of the brightest Hickson Compact Groups contains NGC 5350, 5353, 5354, 5355, 5358. Easy to locate 5′ SE a mag 6.5 star
Seyfert's Sextet	15 59.2	+20 46	14.3	Ser	Seyfert's Sextet is NGC 6027, an extremely compact group! Use high power to resolve three individual members
Hickson 82	16 28.4	+32 51	13.6	Her	Also discovered by Stephan. Hickson 82 consists of the trio NGC 6161, 6162, 6163 within 2.7′. A 4th 16 mag galaxy is close to 6161
Shakhbazian 16	16 49.2	+53 25	14.6	Dra	Five extremely faint mag 15–16 galaxies in a 3.5′ chain oriented N-S. Furthermore, view is severely hampered by a mag 9 star just 2′ NE!
Shakhbazian 166	16 52.8	+81 38	14.9	UMi	Remarkable group of 5 or more extremely faint galaxies in a 7′ chain oriented SW-NE. Located 30′ SE mag 4.2 Epsilon Ursa Minoris
Hickson 86	19 52.1	−30 49	13.1	Sgr	ESO 461–007 is the brightest of four galaxies within 3′. Use high power to resolve
Hickson 90	22 02.1	−31 59	11.4	PsA	Bright quartet containing the interacting system NGC 7174 and 7176, along with NGC 7172 and 7173
Stephan's Quintet	22 36.1	+33 57	12.6	Peg	A 6″ can reveal three members of Stephan's Quintet and a 10″ or 11″ may resolve all five members. Located 30′ SSW of N7331
Hickson 97	23 47.4	02 18	12.9	Psc	Quartet of IC galaxies with brightest member IC 5357. Other members include IC 5351, 5356, 5359

Note: Observation notes by Steve Gottlieb

Table 10.12. Arp interacting pairs (Courtesy of Steve Gottlieb/"Off the Deep End – 10 Challenging Observing Projects")

Object	RA	Dec	Size	V Mag.	Const.	Notes
Arp 144	00 06.4	−13 25	0.9′ × 0.5′	13.9	Cet	Double system containing elongated NGC 7828 with an extremely compact companion, NGC 7829 just off the SE end
Arp 112	00 01.4	+31 26	1.2′ × 0.9′	13.3	Peg	Arp 112 consists of a close interacting pair of small galaxies, NGC 7804 and 7805, less than 1′ between centers
Arp 84	13 58.6	+37 26	2.9′ × 1.5′	11.4	CVn	NGC 5394 and 5395 form a contact pair. Look for a very faint dust lane and arm on the west side of NGC 5395
Arp 199	14 17.0	+36 34	0.9′ × 0.9′	13.4	Boo	Consists of NGC 5545, a very elongated streak SW-NE, with a small knot (NGC 5544) attached at the SW end
Arp 90	15 26.1	+41 41	1.8′ × 0.7′	12.2	Boo	Nearly merged double system with NGC 5929 and 5930 in contact (just 34″ separation between centers)
Arp 91	15 34.5	+15 12	1.7′ × 1.3′	12.1	Ser	Bright double system forms a striking double with NGC 5953 and 5954 attached at the NE edge, just 46″ between centers
Arp 293	16 58.5	+58 56	1.3′ × 1.2′	13.3	Dra	NGC 6286 is the brighter edge-on with a smaller companion, NGC 6285, just 1.5′ NW
Arp 81	18 13.0	+68 21	2.1′ × 0.8′	13.1	Dra	Double galaxy in a common envelope consisting of NGC 6621 and 6622 just 40″ between centers
Arp 93	22 28.6	−24 50	2.3 × 1.4	11.9	Aqr	NGC 7285 is the NE member of a double system with NGC 7284. The close pair is just 30″ between centers!
Arp 284	23 36.2	+02 09	1.9 × 1.4	12.5	Psc	Just 4′ NW of a mag 5.7 star is a 1.8′ pair consisting of brighter NGC 7714 and a faint edge-on, NGC 7715
Arp 86	23 47.1	+29 29	0.8 × 0.5	12	Peg	This is a M51-type system with brighter member NGC 7753 and fainter NGC 7752 2′ SW attached at the end of a spiral arm

Note: Observation notes by Steve Gottlieb

Table 10.13. Abell galaxy clusters (Courtesy of Steve Gottlieb/"Off the Deep End – 10 Challenging Observing Projects")

Name	RA	Dec	Size	V Mag.	Const.	Notes
Abell 262	01 52.7	+ 36 09	120	13.3	And	Relatively loose Abell cluster with over 50 galaxies visible in an 18-in.. Core includes NGC 703, 704, 705, 708, 709, 710
Perseus Galaxy Cluster	03 18.6	+ 41 31	30	12.5	Per	The remarkably rich Perseus galaxy cluster with brightest member NGC 1275 is embedded in a Milky Way field. 300 galaxies tracked down by Albert Highe
Hydra I Cluster	10 36.8	− 27 32			Hya	Hydra I is one of the nearest and easily viewed rich clusters. Core included NGC 3308, 3309, 3311, 3312, 3314, 3316
Leo Galaxy Cluster	11 44.5	+ 19 50	30	13.5	Leo	The Leo cluster is one the richest galaxy clusters for amateurs to explore. Core incudes NGC 3837, 3840, 3841, 3842, 3844, 3845, 3851
Coma I Galaxy Cluster	12 59.8	+ 27 59	120	13.3	Com	The richest galaxy cluster for amateurs. Center on the two brightest members, NGC 4872 and 4889 and then enjoy the view
Cor Bor Galaxy Cluster	15 22.7	+ 27 43	30	15.6	CrB	At 1 billion light years, the most distant cluster visible in an 18-in. scope!
Abell 2147	16 02.3	+ 15 55		13.8	Her	Includes giant galaxies NGC 6146, 6160 and 6173 as well as 6145, 6147, 6150, 6174, 6175, 6180, 6184. Located 2° SW of the Hercules Galaxy Cluster
Hercules Galaxy Cluster	16 05.1	+ 17 45	40	13.8	Her	Rich, irregular cluster, ~500 million l.y. distant, with many spirals and doubles. Core includes NGC 6040, 6041, 6042, 6043, 6045, 6047, 6050, and 6054
Abell 2197	16 28.2	+ 40 54	60	13.9	Her	Includes giant galaxies NGC 6146, 6160 and 6173 as well as 6145, 6147, 6150, 6174, 6175, 6180, 6184. Located 2° SW of the Hercules Galaxy Cluster
Abell 2199	16 28.6	+ 39 31	40	13.9	Her	Swarm of faint galaxies surrounding giant cD galaxy NGC 6166

Note: Observation notes by Steve Gottlieb

Table 10.14. Galaxies Galore!

Table 10.14a. Warm up

Name	Type	RA	Dec.	Mag.	Size	Const.	Notes
NGC 4323	SB0	12 23 1.6	+ 15 54 20	13.5	1.1′ × 0.8′	Vir	
NGC 4232	SBg	12 16 48.7	+ 47 26 18	13.8	1.4′ × 0.7′	CVn	
NGC 4226	Sa/P	12 16 26	+ 47 1 32	13.7	1.3′ × 0.6′	CVn	
NGC 3641	E	11 21 9	+ 3 11 40	13.1	1.1′ × 1.1′	Leo	
NGC 5814	Sab	15 1 21	+ 1 38 14	13.9	0.9′ × 0.5′	Vir	

Seyfert's Sextant Galaxy Group (Source: 2003 issue of *Sky and Telescope*, p. 118), combined magnitude 12.6

(continued)

Suggested Projects by Survey

Table 10.14. (continued)

Table 10.14b. The M49 Group of Galaxies (Source: April 2003 issue of *Sky and Telescope*, pp. 117–119)

Name	Type	RA	Dec.	Mag.	Size	Const.	Notes
NGC 4466	Sab	12 29 31	+07 41 49	15	1.4′	Vir	(Galaxy positions courtesy of www.deepskypedia.com)
NGC 4470	Sa	12 29 38	+07 49 26	12.2	1.3′ × 0.9′	Vir	
PGC 41107		12 28 59	+07 51 04		1.0′ × 0.5′	Vir	
NGC 4434	E	12 27 36.7	+08 15	12.1	1.4′ × 1.4′	Vir	
NGC 4416	SBc	12 26 46.8	+07 55 08	12.4	1.7′ × 1.5′	Vir	
UGC 7580		12 27 55	+08 05	13.3	1.0′ × 0.9′	Vir	
NGC 4366	E?	12 24 45.9	+07 21 09	14.3	0.8′ × 0.4′	Vir	
IC 3259	SBDm	12 23 48.6	+07 11 11	13.6	1.7′ × 0.9′	Vir	
NGC 4341 (=IC 3260)	SAB(s)0/a	12 23 53.5	+07 06 25	13.2	1.6′ × 0.5′	Vir	
IC 3322A	SBc	12 25 42.7	+07 13 01	13.1	3.4′ × 0.4′	Vir	
IC 789	SB0	12 26 20.6	+07 27 36	13.8	1.0′ × 0.6′	Vir	
IC 3322	SBc	12 25 54.1	+07 33 17	13.5	2.3′ × 0.5	Vir	
IC 3267	Sc	12 24 05.5	+07 02 26	13.4	1.2′ × 1.2′	Vir	
NGC 4518	SB0	12 33 11.7	+07 51 05	13.8	1.0′ × 0.4′	Vir	
PGC 41666		12 33 10.4	+07 50 02		0.7′ × 0.3′	Vir	

Table 10.14c. M101 Knots of Stars and Gas (Source: May 2003 issue of *Sky and Telescope*, p. 118)

Name	Type	RA	Dec.	Mag.	Size	Const.	Notes
NGC 5471	–	14 04 29	+54 23 48	–	–	UMa	Star cloud or knot in M101
NGC 5462	–	14 03 52.9	+54 21 54	–	–	UMa	Star cloud or knot in M101
NGC 5461	–	14 03 40.9	+54 19 02	–	–	UMa	Star cloud or knot in M101
NGC 5455	–	14 03 01.2	+54 14 27	–	–	UMa	Star cloud or knot in M101
NGC 5447	–	14 02 29	+54 16 21	–	–	UMa	Star cloud or knot in M101

Table 10.14d Nearby Objects

Name	Type	RA	Dec.	Mag.	Size	Const.	Notes
Andromeda III	dSph	00 35 33.8	+36 29 52	15.0	4.5′ × 3.0′	AND	http://en.wikipedia.org/wiki/Andromeda_III
Andromeda I	E3 pec?	00 45 39.8	+38 02 28	13.6	2.5′ × 2.5′	AND	http://en.wikipedia.org/wiki/Andromeda_I
Fornax Dwarf[a]	dE0	02 39 59.3	−34 26 57	9.3	17.0′ × 12.6′	FOR	http://en.wikipedia.org/wiki/Fornax_Dwarf
Andromeda II	dSph	01 16 29.8	+33 25 09	13.5	3.6′ × 2.5′	AND	http://en.wikipedia.org/wiki/Andromeda_II
Maffei 1	E3	02 36 35.4	+59 39 19	11.4	3.4′ × 1.7′	CAS	http://en.wikipedia.org/wiki/Maffei_1

[a]Has at least 3 globular clusters visible: NGC 1049, Fornax 1, and Fornax 2

Table 10.14e. Pegasus Groups 1 and 2

Name	Type	RA	Dec.	Mag.	Size	Const.	Notes
NGC 7465	SB0	23 02 00.9	+15 57 53	12.6	1.2′ × 0.7′	Peg	Group 1
NGC 7464	E1	23 01 53.6	+15 58 26	13.3	0.5′	Peg	Group 1
NGC 7463	SBb/P	23 01 51.9	+15 58 55	12.7	2.9′ × 0.7′	Peg	Group 1
NGC 7385	E1	22 49 54.6	+11 36 30	13.0	1.5′ × 1.3′	Peg	Group 2
NGC 7386	E-S0	22 50 02.1	+11 41 54	12.3	1.4′ × 1.2′	Peg	Group 2
NGC 7384	(star)	22 49 42.5	+11 29 17	14.7	–	Peg	Group 2
Pegasus I		23 20.5	+8 11			Peg	

(continued)

Table 10.14. (continued)

Table 10.14f. Behind M44 (M44 is 570 LY away, the galaxies are 200 MLY away)

Name	Type	RA	Dec.	Mag.	Size	Const.	Notes
NGC 2624	S	08 38 09.5	+19 43 31	13.9	0.8′ × 0.7′	Cnc	
NGC 2625	E	08 38 23.1	+19 42 56	14.3	0.3′ × 0.3′	Cnc	

Table 10.14g. Group: Abell 2572

Name	Type	RA	Dec.	Mag.	Size	Const.	Notes
NGC 7592	S0-a	23 18 22.1	−04 25 01	13.5	1.2′ × 0.9′	Aqr	
NGC 7602	S0-a	23 18 43.3	+18 41 52	14.4	0.5′ × 0.5′	Peg	
NGC 7598	Pec.	23 18 33.3	+18 45 0	14.9	0.3′ × 0.2′	Peg	
NGC 7588	S	23 17 57.7	+18 45 07	14.9	0.3′ × 0.2′	Peg	
NGC 7578A	S0	23 17 13.5	+18 42 30	13.9	23″ × 18″	Peg	Otherwise known as Arp 170, coordinates are for center of pair
NGC 7578B	E1	23 17 13.5	+18 42 30	14.7	52″ × 44″	Peg	Otherwise known as Arp 170, coordinates are for center of pair

Table 10.14h. Group: NGC 507 (Source: Dec. 2003 issue of *Sky and Telescope*, pp. 119–120)

Name	Type	RA	Dec.	Mag.	Size	Const.
NGC 494	Sab	01 22 55.4	+33 10 27	13	2.0′ × 0.8′	Psc
NGC 504	S0	01 23 27.9	+33 12 17	13.1	1.7′ × 0.4′	Psc
NGC 507	SA(R)0	01 23 40.0	+33 15 22	11.7	3.1′ × 3.1′	Psc
NGC 495	SB0/Sba	01 22 56.0	+33 28 18	13.1	1.3′ × 0.8′	Psc
NGC 499	S0-	01 23 11.5	+33 27 37	12	1.6′ × 1.3′	Psc
NGC 508	E	01 23 40.5	+33 16 50	13.2	1.3′ × 1.3′	Psc
PGC 5129	S0	01 23 58.5	+33 18 48	15.5B	0.4′ × 0.3′	Psc
IC 1687	C	01 23 19.3	+33 16 37	13.7	0.5′ × 0.3′	Psc
NGC 503	E	01 23 28.5	+33 19 56	14.1		Psc
IC 1690	S0	01 23 49	+33 9 25	14.4		Psc
IC 1689	S0	01 23 49.5	+33 09 25	14.4	0.5′ × 0.3′	Psc
NGC 504	S0	01 23 27.9	+33 12 17	13.1	1.7′ × 0.4′	Psc
IC 1685	S	01 23 06.7	+33 11 27	14.7	0.5′ × 0.4′	Psc
IC 1682	S?	01 22 13.4	+33 15 35	13.6	0.8′ × 0.6	Psc
IC 1673	E	01 20 46.3	+33 02 44	14.1	0.3′ × 0.3′	Psc
IC 1677	S?	01 21 07.1	+33 12 56	14.1	0.9′ × 0.5′	Psc
NGC 496	SB-c	01 23 11.7	+33 31 48	13.5	1.6′ × 0.9′	Psc
NGC 498	S0	01 23 11.3	+33 29 22	14.3	0.3′ × 0.3′	Psc
NGC 501	E0	01 23 22.3	+33 25 59	14.5		Psc
NGC 483	S	01 21 56.3	+33 31 17	13.2	0.7′ × 0.7′	Psc
IC 1679	S?	01 21 44.5	+33 29 34	14.8	0.5′ × 0.4′	Psc
NGC 515	S0	01 24 38.7	+33 28 20	13.1	1.4′ × 1.1′	Psc
NGC 517	S0	01 24 43.8	+33 25 44	12.5	1.4′ × 0.5′	Psc
NGC 512	Sa-b	01 23 59.8	+33 54 29	13.4	1.6′ × 0.4′	And
NGC 513	S	01 24 26.8	+33 47 58	12.9	0.7′ × 0.3′	And
NGC 523	Pec.	01 25 20.8	+34 01 31	13	2.5′ × 0.8′	And
NGC 528	S0	01 25 33.6	+33 40 17	12.6	1.7′ × 1.1′	And

(continued)

Table 10.14. (continued)

Table 10.14i Abell 2199 (Source: Jul. 2009 issue of *Sky and Telescope,* pp. 61–62)

Name	Type	RA	Dec.	Mag.	Size	Const.
NGC 6166	E	16 28 38.5	+39 33 06	13.5	1.9′ × 1.4′	Her
NGC 6166A	S0	16 28 31	+39 31 16	15		Her
NGC 6166D	S0	16 28 39	″ + 39 31 8	14.8		Her
NGC 6166C	E	16 28 23	+39 34 14	14.5		Her
NGC 6166B	S?	16 28 53	+39 33 39	14.9		Her
NGC 6158	E-S0	16 27 40.9	+39 22 59	13.8		Her

Table 10.14j. NGC 80 Group (Source: Nov. 2009 *Sky and Telescope,* pp. 66–67)

Name	Type	RA	Dec.	Mag.	Size	Const.
NGC 80	SA0-	00 21 10.9	+22 21 26	12.5	1.8′ × 1.7′	And
NGC 81	S	00 21 13.3	+22 22 58	15.7	0.2′ × 0.1′	And
NGC 68	SA0-	00 18 18.2	+30 04 19	13.9	2.0′ × 1.8′	And
NGC 71	SA0- pec	00 18 18.2	+30 04 19	14.4	1.2; × 1.1′	And
NGC 67	E	00 18 18.2	+30 04 19	15.2	1.0′ × 0.7′	And
NGC 69	SB0(s)	00 18 18.2	+30 04 19	15.3	0.9′ × 0.8′	And
NGC 72A (=PGC 1208)	E3	18 34 00	+30 02 10	14.7		And
NGC 74	Sb	00 18 49.3	+30 03 41	15.3	0.8′ × 0.3′	And
NGC 76	S?	00 19 37.7	+29 56 01	13	1.0′ × 0.9′	And

Table 10.15. Hickson compact groups (Courtesy of Steve Gottlieb/"Off the Deep End – 10 Challenging Observing Projects")

HCG #	Other	PGC	Type	RA	Dec.	Size	PA	V Mag	Surf Brt	Radial Vel
001A	UGC 00248	1627	S?	00 26 07.1	+25 43 30	1.7′ × 1.1′		13.9	14.4	9,913 km/s
001B		1625	S0?	00 26 06.1	+25 43 08			14.5		9,937
001 C		1614	S?	00 25 54.4	+25 43 25			14.9		9,732
002A	UGC 00312	1921	SB?	00 31 23.9	+08 28 01	1.4 × 0.7	70	12.9	12.7	3,992
002B	MCG +01-02-018	1914	S?	00 31 18.9	+08 28 29	0.7 × 0.5		13.5	12.2	4,015
002C	UGC 00314	1927	S?	00 31 29.4	+08 24 02	1.1 × 0.9	165	14	13.8	3,932
002D	UGC 00315	1934	Sbc	00 31 38.4	+08 23 26	0.9 × 0.9				3,933
003A		2045	S?	00 34 13.2	−07 33 54			14.5		6,972
003B		2064	S0?	00 34 25.1	−07 35 59			14.6		7,530
003D		2043	E?	00 34 09.8	−07 36 08			14.8		7,474
004A	ESO 540-001	2047	(R')SB(r)bc	00 34 13.7	−21 26 20	1.4 × 1.0	21	13	13.1	7,775
004B	ESO 540-002	2046	(R')SB(r)b: pec	00 34 14.0	−21 28 16	1.1 × 0.4	122	14.8	13.8	6,764
004C		2051	S0?	00 34 15.5	−21 25 04			14.8		8,562
004D		2057	E?	00 34 16.8	−21 28 40			14.8		7,914
004E		2040	S?	00 34 08.4	−21 26 09			15.5		
005A	NGC 0190n	2324	Sab	00 38 54.7	+07 03 46	0.9 × 0.8		14	13.5	10,857
005B	NGC 0190s	2325	S0?	00 38 54.7	+07 03 24	0.3 × 0.3		14.7	11.6	11,777
005C		2322	S0	00 38 52.8	+07 04 22	0.25 × 0.2				
005D		2326	S	00 38 54.9	+07 02 53	0.4 × 0.15		17.3		
006A	NPM1G −08.0024	2353	S0?	00 39 13.8	−08 23 39			14.7		11,342
006B		2350	S?	00 39 10.2	−08 23 50			15		
006C		2351	E?	00 39 11.1	−08 24 01			14.7		
007A	NGC 0192	2352	SBa	00 39 13.4	+00 51 49	1.9 × 0.9	167	12.6	13.1	

(continued)

Table 10.15. (continued)

HCG #	Other	PGC	Type	RA	Dec.	Size	PA	V Mag	Surf Brt	Radial Vel
007B	NGC 0196	2357	SB0 pec	00 39 17.8	+ 00 54 46	1.3 × 0.8	15	12.9	12.8	3,902
007C	NGC 0201	2388	Sc	00 39 34.9	+ 00 51 35	1.8 × 1.4	155	12.9	13.6	
007D	NGC 0197	2365	SB0° pec	00 39 18.8	+ 00 53 31	1.2 × 0.8	5	14.1	14	3,856
008A	MCG + 04-03-008	2886	S?	00 49 34.1	+ 23 34 42	0.9 × 0.7		14.2	13.5	15,697
008B	MCG + 04-03-007	2888	S?	00 49 35.4	+ 23 35 28	0.3 × 0.1		15.2	11.5	15,649
008C	MCG + 04-03-009	2890	S?	00 49 36.0	+ 23 35 02	0.2 × 0.2		15.1	11.4	16,770
008D		2892	S?	00 49 36.9	+ 23 34 22			15.2		16,024
009A	MCG − 04-03-028	3201	S?	00 54 20.5	− 23 33 09	0.7 × 0.4		14.3	12.8	19,870
009B		3196	S0?	00 54 16.2	− 23 32 08			14.8		
010A	NGC 0536	5344	SB(r)b	01 26 21.6	+ 34 42 14	3.0 × 1.1	62	12.4	13.5	4,921
010B	NGC 0529	5299	S0⁻:	01 25 40.3	+ 34 42 47	2.4 × 2.1	160	12.1	13.8	4,529
010C	NGC 0531	5340	SB0/a:	01 26 18.9	+ 34 45 16	1.9 × 0.5	34	13.8	13.6	4,915
010D	NGC 0542	5360	S?	01 26 30.8	+ 34 40 32	1.0 × 0.2		14.7	12.9	4,390
011A	ESO 476-008	5362	SAB(rs)c	01 26 33.3	− 23 13 33	1.9 × 1.8		12.7	13.9	5,240
011B	MCG − 04-04-010	5365	S?	01 26 34.0	− 23 15 56	0.4 × 0.2		14.8	12.1	7,088
012A	MCG − 01-04-052	5437	E?	01 27 33.4	− 04 40 58			14.1		14,109
013A	MCG − 01-05-002	5732	SA(s)b:	01 32 20.5	− 07 51 34	1.0 × 0.5		14.8	13.8	12,179
013B	MCG − 01-05-003	5735	S?	01 32 22.5	− 07 52 27			14.7		
014A	MCG − 01-06-020	7553	SA0° pec	01 59 49.9	− 07 03 34	1.6 × 0.6		14.5	14.4	5,664
014B	MCG − 01-06-022	7557	SAb pec?	01 59 52.2	− 07 05 12	1.1 × 0.4		14.1	13	
015A	UGC 01624	8128	S0:	02 07 53.1	+ 02 10 03	1.1 × 0.5	130	13.7	13	6,766
015B	UGC 01617	8110	S0⁻:	02 07 34.2	+ 02 06 55	0.8 × 0.7		14	13.2	6,874
015C	UGC 01620	8117	S0⁻:	02 07 39.8	+ 02 08 59	0.8 × 0.8		13.6	13	6,991
015D	UGC 01618	8114	S0⁻:	02 07 37.6	+ 02 10 51	0.9 × 0.8		14.3	13.7	6,008
015E	MCG + 00-06-033	8096	S0?	02 07 25.4	+ 02 06 58	0.3 × 0.3		14.8	11.8	6,932
016A	NGC 0835	8228	SAB(r)ab: pec	02 09 24.9	− 10 08 09	1.3 × 1.0		12.1	12.3	3,867
016B	NGC 0833	8225	(R')Sa: pec	02 09 21.1	− 10 08 00	1.5 × 0.7	75	12.7	12.6	3,682
016C	NGC 0838	8250	SA(rs)0° pec:	02 09 38.5	− 10 08 47	1.1 × 0.9		13	12.8	3,590
016D	NGC 0839	8254	S pec sp	02 09 43.0	− 10 11 03	1.4 × 0.7	75	13.1	13	3,583
017A		8561		02 14 05.0	+ 13 04 41					
018A	UGC 02140A	10046	SB0/a: sp	02 39 09.4	+ 18 22 03	0.9 × 0.2		14.5	12.3	9,791
018B	UGC 02140	10044	IB(s)m pec	02 39 06.3	+ 18 22 58	1.7 × 0.7	155	14.9	15	3,848
018C		10043	S?	02 39 06.0	+ 18 23 19			15.1		3,915
018D		10042	S?	02 39 04.8	+ 18 23 38			14.6		3,839
019A	MCG − 02-07-073	10262	S0⁻ pec	02 42 38.3	− 12 25 19	1.2 × 0.9	40	13.2	13.1	4,077
019B	MCG − 02-07-074	10268	(R)SB(r)a pec?	02 42 42.2	− 12 25 42	0.9 × 0.3	100	14.8	13.2	4,049
019C	MCG − 02-07-075	10270	SBm pec?	02 42 46.4	− 12 23 54	1.3 × 0.7	100	14.9	14.7	3,949
020A		10364		02 44 12.0	+ 26 05 57					
021A	NGC 1099	10422	SB(rs)b	02 45 17.7	− 17 42 30	1.8 × 0.6	10	13.1	13	7,414
021B	NGC 1100	10438	SAB(r)a:	02 45 35.7	− 17 41 13	1.7 × 0.7	58	13	13.1	7,356
021C	NGC 1098	10403	SA(s)0⁻:	02 44 53.8	− 17 39 35	1.8 × 1.3	102	12.6	13.4	7,187
021D	NGC 1092	10432	E?	02 45 29.8	− 17 32 31	0.9 × 0.8	170	13.4	13.1	8,636
021E	NGC 1091	10424	(R')SAB(rs)a pec?	02 45 21.8	− 17 31 54	0.9 × 0.6	77	14.1	13.3	
022A	NGC 1199	11527	E3	03 03 38.5	− 15 36 50	2.4 × 1.9		11.4	13	2,488
022B	NGC 1190	11508	S0°: sp	03 03 26.1	− 15 39 44	0.9 × 0.3		14.2	12.6	2,444
022C	NGC 1189	11503	SB(s)dm	03 03 24.3	− 15 37 25	1.7 × 1.5		13.9	14.8	2,547
022D	NGC 1191	11514	SB0⁻	03 03 30.9	− 15 41 08	0.6 × 0.5		14.3	12.7	9,162
022E	NGC 1192	11519	E?	03 03 34.7	− 15 40 45	0.7 × 0.3		14.4	12.7	9,326
023A	NGC 1214	11675	SB(rs)0⁺?	03 06 56.0	− 09 32 37	1.3 × 0.3		14	12.8	4,657
023B	NGC 1215	11687	(R)SAB(r)a pec	03 07 09.6	− 09 35 32	1.5 × 1.1	15	14.1	14.5	4,749

(continued)

Table 10.15. (continued)

HCG #	Other	PGC	Type	RA	Dec.	Size	PA	V Mag	Surf Brt	Radial Vel
023C	NGC 1216	11693	S0$^+$? sp	03 07 18.6	−09 36 43	0.8 × 0.2	65	14.8	12.8	4,832
023D	MCG −02-08-050	11673		03 06 55.2	−09 37 46	0.7 × 0.5		15.3	16.3	
024A	MCG −02-09-031	12477	S?	03 20 15.4	−10 51 48	0.5 × 0.4		14.7	13.1	9,083
024B	MCG −02-09-032	12501	S?	03 20 22.9	−10 52 03	0.4 × 0.4		14.5	12.5	8,972
025A	UGC 02690	12531	SAc pec:	03 20 42.9	−01 06 29	1.3 × 0.6	143	13.9	13.5	
025B	UGC 02691	12539	SBa	03 20 45.4	−01 02 42	1.0 × 0.6		14	13.2	6,227
025C	CGCG 390-070	12533	S?	03 20 43.4	−01 00 09			15		6,111
025D	CGCG 390-067	12524	S?	03 20 38.9	−01 02 07			15.2		6,111
025F		12538	S0?	03 20 45.6	−01 03 15			15.5		
026A	MCG −02-09-035	12611	SBd pec sp	03 21 55.4	−13 39 03	1.1 × 0.2				9,518
026B		12614	S0?	03 21 57.2	−13 38 55			15.4		9,173
027A		14873		04 19 26.9	−11 44 01					
027B		14863	S?	04 19 18.7	−11 42 07			15.4		
028A		15136	S?	04 27 18.6	−10 18 22			15.2		11,224
028B		15141	S?	04 27 20.1	−10 19 35			14.9		11,332
028C		15135	E?	04 27 18.3	−10 19 05			15.5		
029A		15559	S?	04 34 43.3	−30 32 39			14.1		13,289
030A	MCG +00-12-051	15620	(R')SB(rs)0$^+$:	04 36 18.6	−02 49 53	1.2 × 0.8		12.6	12.4	4,635
030B	MCG +00-12-054	15631	(R')SAB(rs)0$^+$:	04 36 30.2	−02 51 59	1.1 × 0.9		13.2	13.1	4,563
030C	PGC 15624	15624	S?	04 36 23.3	−02 48 00			14.7		4,446
030D	PGC 15636	15636		04 36 36.5	−02 50 36					
031A	NGC 1741	16574	Pec	05 01 38.4	−04 15 25	1.4 × 0.7		12.5	12.4	4,028
031B		16570	SB(s)m pec sp	05 01 35.5	−04 15 51			14.7		4,095
031C		16573	SB(s)m pec:	05 01 37.9	−04 15 28			12.9		4,046
031D		16571	S?	05 01 35.5	−04 15 24					
032A	MCG −03-13-053	16583	S0?	05 01 45.2	−15 26 56			13.4		12,533
032B		16578	S?	05 01 40.2	−15 23 54			14.9		12,111
033A	CGCG 469-002	16867		05 10 47.7	+18 01 11			14.2		
033B	CGCG 469-002n	16866		05 10 47.3	+18 01 49					
034A	NGC 1875	17171	S0?	05 21 46.0	+06 41 20	1.6 × 0.4		13.6	13.1	8,999
034B		17176	SA0$^-$:	05 21 50.0	+06 40 36					9,516
035A	MCG +08-16-028	24601	S?	08 45 21.3	+44 31 14	0.3 × 0.1		15.1	11.4	16,094
035B		24597	S0?	08 45 20.6	+44 30 32			14.5		16,513
035C		24596	S0?	08 45 18.5	+44 31 40			15		16,532
036A	IC0528	25783	Sab	09 09 22.6	+15 47 46	1.5 × 0.8	163	14.1	14.1	4,065
037A	NGC 2783	26013	E	09 13 39.5	+29 59 34	2.1 × 1.5	168	12.6	13.9	6,944
037B	UGC 04856	26012	Sb	09 13 32.9	+29 59 59	1.9 × 0.2	77	14.3	13.3	6,969
037C	IC2449:	26009		09 13 37.4	+29 59 58	0.3 × 0.2				
037D	MCG +05-22-016	26005	S0?	09 13 33.9	+30 00 52			15.4		7,604
037E	MCG +05-22-018	26006	S?	09 13 34.0	+30 02 23			15.3		6,978
038A	MCG +02-24-012	26831	S?	09 27 34.7	+12 16 10	0.8 × 0.2		15	12.7	9,063
038B	MCG +02-24-013	26842	S?	09 27 43.6	+12 17 14			14.3		8,994
038C	MCG +02-24-014	26844	S?	09 27 44.6	+12 17 17	0.4 × 0.2		15	12.1	9,038
039A	UGC 05057a	26926		09 29 27.7	−01 20 44					
040A	MCG −01-25-009	27509	E	09 38 53.4	−04 50 56	1.3 × 1.0		12.8	13.1	6,950
040B	MCG −01-25-010	27513	SA(r)0$^-$: pec	09 38 54.9	−04 51 56	1.1 × 0.7		14	13.6	7,149
040C	MCG −01-25-008	27508	SB(rs)b pec sp	09 38 53.5	−04 51 37	1.1 × 0.3		14.9	13.6	7,180
040D	MCG −01-25-012	27516	SB(s)0/a pec:	09 38 55.8	−04 50 14	0.9 × 0.4		14.2	12.9	6,794
040E	MCG −01-25-011	27515	SAB(s)a pec:	09 38 55.5	−04 51 28	0.7 × 0.3				6,952
041A	UGC 05345	28764	S?	09 57 35.6	+45 13 48	1.5 × 0.3	65	13.6	12.4	3,962
041B	UGC 05346	28770	S?	09 57 40.8	+45 15 32	0.9 × 0.3	28	14.3	12.7	7,452

(continued)

Table 10.15. (continued)

HCG #	Other	PGC	Type	RA	Dec.	Size	PA	V Mag	Surf Brt	Radial Vel
041C	MCG + 08-18-046	28753	S?	09 57 27.9	+ 45 14 18	0.4 × 0.1		15.5	12.1	9,928
042A	NGC 3091	28927	E3:	10 00 13.9	− 19 38 14	3.0 × 1.9	149	11.1	13	4,209
042B	NGC 3096	28950	SB(rs)0°	10 00 32.9	− 19 39 45	1.0 × 0.8	170	13.4	13	4,529
042C	MCG − 03-26-006	28922	E?	10 00 10.3	− 19 37 19	0.4 × 0.4		13.2	11.1	4,336
042D		28926	S0?	10 00 13.0	− 19 40 22			15.2		4,407
043A	CGCG 008-062	29677	S?	10 11 19.9	− 00 01 23			14.8		10,493
043B	CGCG 008-059	29657	S?	10 11 07.7	− 00 02 33			15.1		10,417
043C	CGCG 008-061	29665	S0?	10 11 12.7	− 00 04 04			15.1		10,259
044A	NGC 3190	30083	SA(s)a pec sp	10 18 05.7	+ 21 49 57	4.4 × 1.5	125	11.1	13.1	1,641
044B	NGC 3193	30099	E2	10 18 25.0	+ 21 53 37	3.0 × 2.7		10.9	13.1	1,687
044C	NGC 3185	30059	(R)SB(r)a	10 17 38.5	+ 21 41 18	2.3 × 1.6	130	12.2	13.4	1,539
044D	NGC 3187	30068	SB(s)c pec	10 17 47.8	+ 21 52 25	3.0 × 1.3		13.4	14.7	1,886
045A	UGC 05564	30153	Sb	10 19 13.8	+ 59 07 51	1.3 × 0.4	75	14.9	14.1	21,957
046A	MCG + 03-27-005	30347	S0?	10 22 07.3	+ 17 50 17	0.5 × 0.4		14.5	12.6	8,521
046B	MCG + 03-27-007	30349	S0?	10 22 12.9	+ 17 51 13			15.5		8,226
046D	MCG + 03-27-009	30354		10 22 17.2	+ 17 52 54					
047A	UGC 05644	30616	SA(r):	10 25 46.3	+ 13 43 00	1.0 × 0.6	15	14.2	13.5	9,911
047B	MCG + 02-27-013	30619	S0?	10 25 48.7	+ 13 43 40	0.8 × 0.8	15		14.3	9,817
048A	IC2597	31586	E+4	10 37 47.7	− 27 04 55	2.6 × 1.8	4	11.8	13.5	3,338
048B	ESO 501-059	31588	S?	10 37 49.7	− 27 07 19	0.9 × 0.8	97	14	13.5	2,716
048C		31577	S?	10 37 40.6	− 27 03 28			15.4		4,534
049A	CGCG 314-001	32899	S?	10 56 41.6	+ 67 11 07			15.2		10,043
050A		34447		11 17 06.4	+ 54 55 00					
051A	NGC 3651	34898	E	11 22 26.5	+ 24 17 56	1.1 × 1.1		13.2	13.4	8,007
051B	MCG + 04-27-026	34882	S?	11 22 14.2	+ 24 18 00.5	0.8 × 0.6		14.6	13.5	8,494
051C	NGC 3653	34905	S0?	11 22 30.3	+ 24 16 45	0.9 × 0.6		13.6	12.8	9,213
051D	MCG + 04-27-030	34907	S?	11 22 30.6	+ 24 18 00	0.3 × 0.3		14.7	11.7	7,840
051E	IC2759	34881	S0?	11 22 13.3	+ 24 19 02	0.3 × 0.3		14.1	11	8,011
051F		34899	S0?	11 22 26.7	+ 24 17 37			14.2		7,843
051G		34901	S0?	11 22 28.4	+ 24 17 42			15.2		7,843
052A	MCG + 04-27-036	35183	SB?	11 26 19.1	+ 21 05 46	0.7 × 0.3		14.4	12.6	13,300
053A	NGC 3697	35347	SABb	11 28 50.4	+ 20 47 43	2.3 × 0.7	93	13.1	13.5	6,584
053B	MCG + 04-27-044	35360	S0?	11 28 59.9	+ 20 44 21	0.5 × 0.4		13.9	12.1	
053C	MCG + 04-27-045	35355	S?	11 28 58.5	+ 20 44 59	0.7 × 0.5		14.1	12.9	
054A	IC0700	35382	S0?	11 29 15.3	+ 20 35 00	1.0 × 0.5		13	12.1	1,737
054B		35380	S?	11 29 14.1	+ 20 34 53			15.2		1,734
055A	MCG + 12-11-028A	35575	SA0 pec:	11 32 07.0	+ 70 48 56			14.9		15,732
055B	MCG + 12-11-028B	35572	E pec:	11 32 05.5	+ 70 48 24					15,563
055C	MCG + 12-11-028C	35573	SBa pec:	11 32 05.7	+ 70 48 40					15,654
055D	MCG + 12-11-028D	35574	S0 pec:	11 32 06.9	+ 70 49 17					15,973
055E	MCG + 12-11-028E	35576	S?	11 32 07.6	+ 70 49 08					36,963
056A	MCG + 09-19-113	35631	S?	11 32 46.7	+ 52 56 27	1.1 × 0.2		15.2	13.3	8,431
056B	UGC 06527	35620	Sa pec sp	11 32 38.9	+ 52 56 53	1.1 × 0.3		15.3	14.1	8,301
056C		35618	(R')S0/a pec:	11 32 36.7	+ 52 56 51			14.8		8,296
056D		35615	SA(s)0/a pec:	11 32 35.4	+ 52 56 50					8,532
056E		35609	SB0 pec:	11 32 32.8	+ 52 56 22			15.4		8,070
057A	NGC 3753	36016	Sab? pec sp	11 37 53.8	+ 21 58 53	1.7 × 0.5	120	13.6	13.2	9,034
057B	NGC 3746	35997	SB(r)b	11 37 43.6	+ 22 00 35	1.1 × 0.5	127	14.2	13.5	9,342
057C	NGC 3750	36011	SAB0⁻?	11 37 51.7	+ 21 58 27	0.8 × 0.7		13.9	13.1	9,381
057D	NGC 3754	36018	SBb? pec	11 37 55.0	+ 21 59 07	0.4 × 0.3		14.3	12	9,329
057E	NGC 3748	36007	SB0°? sp	11 37 49.1	+ 22 01 34	0.7 × 0.4		14.8	13.2	9,306

(continued)

Table 10.15. (continued)

HCG #	Other	PGC	Type	RA	Dec.	Size	PA	V Mag	Surf Brt	Radial Vel
057F	NGC 3751	36017	S0$^-$ pec?	11 37 53.9	+21 56 11	0.8 × 0.5	5	13.9	12.7	9,909
057G	NGC 3745	36001	SB(s)0$^-$:	11 37 44.4	+22 01 16	0.4 × 0.2		15.2	12.4	9,730
058A	NGC 3822	36319	S0?	11 42 11.3	+10 16 40	1.4 × 0.8	178	13.1	13.1	6,476
058B	NGC 3825	36348	SBa	11 42 23.7	+10 15 52	1.3 × 1.0	160	13	13.2	6,853
058C	NGC 3817	36299	SB0/a	11 41 53.1	+10 18 07	1.0 × 0.9	140	13.3	13.1	6,446
058D	NGC 3819	36311	E?	11 42 06.0	+10 21 03	0.8 × 0.7		13.8	13	6,614
058E	NGC 3820	36308	S?	11 42 04.9	+10 23 02	0.7 × 0.4		14.5	13	6,396
059A	IC0737	36861	E?	11 48 27.5	+12 43 39			13.8		4,452
059B	IC0736	36853	S0?	11 48 20.2	+12 42 58	0.5 × 0.5		14.6	13.1	4,246
059C	MCG +02-30-041	36871	S?	11 48 32.5	+12 42 19	0.9 × 0.2		14.9	13	
059D	MCG +02-30-040	36867	S?	11 48 30.8	+12 43 47	0.7 × 0.4		15.2	13.7	4,425
060A	MCG +09-20-071	38065	S0?	12 03 07.3	+51 40 31			14.8		19,153
061A	NGC 4169	38892	S0	12 12 19.0	+29 10 49	1.8 × 0.9	153	12.2	12.6	4,093
061B	NGC 4173	38897	SBd:	12 12 21.6	+29 12 24	5.0 × 0.7	134	13	14.2	1,391
061C	NGC 4175	38912	S	12 12 30.7	+29 10 09	1.8 × 0.4	130	13.2	12.6	4,224
061D	NGC 4174	38906	S?	12 12 27.0	+29 08 57	0.8 × 0.3	50	13.3	11.7	4,266
062A	NGC 4759e	43757	S0$^+$	12 53 05.8	−09 12 15			12.5		4,684
062B	NGC 4759w	43754	S0° pec	12 53 04.6	−09 12 01			13		3,980
062C	NGC 4761	43768	E$^+$	12 53 09.9	−09 11 53	0.4 × 0.3		13.8	11.6	4,688
062D	NGC 4764	43760	S?	12 53 06.8	−09 15 29			15		
063A	ESO 443-037	44984	(R')SB(s)a	13 02 17.0	−32 45 37	1.1 × 0.5	178	14	13.2	5,518
063B	ESO 381-050	44965	SB(r)bc pec	13 02 07.0	−32 47 13	1.2 × 1.1		13.6	13.9	9,636
063C		44979	S?	13 02 13.2	−32 46 02			14.5		
064A		46975	S?	13 25 45.3	−03 51 50	1.1 × 0.5		14.4	13.7	10,901
064C		46977	S?	13 25 39.5	−03 48 53			14.3		
065A	ESO 444-055	47397	S0?	13 29 51.0	−29 30 53	1.4 × 0.8	29	13.3	13.4	14,412
065B		47406	S0?	13 29 54.9	−29 29 50			14.3		14,989
065C		47403	S?	13 29 53.3	−29 29 31			14.5		14,413
065D		47401	E?	13 29 52.1	−29 30 47			14.6		14,012
066A	MCG +10-19-104	48226	S0?	13 38 38.9	+57 18 44	0.4 × 0.3		14.8	12.5	20,815
067A	NGC 5306	49039	S0 pec?	13 49 11.3	−07 13 28	1.4 × 1.1		12.1	12.4	7,541
067B	MCG −01-35-013	49017	Sb	13 48 59.5	−07 11 47	2.1 × 0.4		14.3	14.1	7,929
067C	MCG −01-35-015	49040	S?	13 49 12.6	−07 12 34	0.7 × 0.3		14.7	12.9	7,653
067D		49036	S0 pec?	13 49 09.9	−07 13 54			14.8		7,356
068A	NGC 5353	49356	S0	13 53 26.7	+40 16 59	2.2 × 1.1	145	11	11.8	2,296
068B	NGC 5354	49354	S0	13 53 26.7	+40 18 10	1.4 × 1.3		11.4	11.9	2,647
068C	NGC 5350	49347	SB(r)b	13 53 21.5	+40 21 49	3.2 × 2.3	40	11.3	13.3	2,492
068D	NGC 5355	49380	S0?	13 53 45.6	+40 20 19	1.2 × 0.7	35	13.1	12.8	2,602
068E	NGC 5358	49389	S0/a	13 54 00.4	+40 16 38	1.1 × 0.3	138	13.6	12.3	2,620
069A	UGC 08842	49502	S?	13 55 29.9	+25 04 26	1.3 × 0.5		14.8	14.1	9,083
069B	MCG +04-33-028	49505	S0?	13 55 34.4	+25 02 58			14.8		8,781
069C	UGC 08842c	49505		13 55 32.6	+25 04 28	0.3 × 0.2				
070A	UGC 08990	50139	Sa	14 04 10.4	+33 20 14	1.2 × 0.2	142	15.2	13.8	8,441
070B	IC4371	50140	S0?	14 04 11.0	+33 18 29	0.9 × 0.6		14.1	13.3	8,401
070C	MCG +06-31-065	50159	S?	14 04 20.8	+33 19 16	0.9 × 0.2		14.2	12	8,282
070D	IC4370			14 04 09.9	+33 20 45	0.4 × 0.4				
070E	IC4369	50134	S?	14 04 06.1	+33 19 13	0.4 × 0.4		15.2	13.2	19,320
070G		50123	Sa	14 04 00.4	+33 19 52			15.4		
071A	IC4381	50629	Scd:	14 10 57.1	+25 29 48	2.5 × 1.4	135	13.7	14.9	9,533
071B	IC4382	50635	S?	14 11 02.6	+25 31 11	0.7 × 0.2		14.4	11.9	9,374
071C		50640	S?	14 11 05.2	+25 28 58			15.3		

(continued)

Table 10.15. (continued)

HCG #	Other	PGC	Type	RA	Dec.	Size	PA	V Mag	Surf Brt	Radial Vel
072A	MCG +03-38-017	52844	S0?	14 47 53.6	+19 04 36	0.5 × 0.1		15.1	12.2	12,700
072B	MCG +03-38-021	52851	S0?	14 47 54.8	+19 03 36	0.6 × 0.2		15.1	12.8	12,752
072C	MCG +03-38-022	52854	S0?	14 47 57.0	+19 02 41	0.3 × 0.3		14.8	12	13,256
072D	MCG +03-38-020	52848	S?	14 47 55.8	+19 03 24	0.2 × 0.2		15	11.2	12,550
073A	NGC 5829	53709	SA(s)c	15 02 42.3	+23 19 56	1.8 × 1.5		13.4	14.3	5,865
073B	IC4526	53707		15 02 38.2	+23 21 02					
074A	NGC 5910	54689	E?	15 19 24.8	+20 53 47	0.9 × 0.8		13.6	13.1	12,407
074B		54688	E?	15 19 24.3	+20 53 26			14.7		12,262
075A	CGCG 135-050	54804	E?	15 21 30.4	+21 11 27			14		12,687
075C		54827	S?	15 21 38.9	+21 10 37			15.5		12,440
075D		54824		15 21 37.3	+21 10 53					
075E		54818		15 21 34.1	+21 10 43					
076A	NGC 5944	55321	S?	15 31 47.7	+07 18 27	0.7 × 0.2		14.9	12.4	10,207
076B	MCG +01-40-003	55314	S0?	15 31 40.4	+07 20 19			13.9		10,156
076C	NGC 5941	55309	S0?	15 31 37.1	+07 18 43			14.3		10,817
076D	NGC 5942	55316	E?	15 31 42.2	+07 17 14			14.8		10,304
077A	UGC 10049	56123	S0?	15 49 16.9	+21 49 10	1.4 × 0.6		13.7	13.3	10,619
077B	UGC 10049	56122	S0?	15 49 16.8	+21 49 25			15.2		
077C	UGC 10049	56121	S?	15 49 16.9	+21 49 51			15.4		
078A	UGC 10057	56079	(R')SB(s)bc	15 48 17.3	+68 13 14	1.4 × 0.7	92	14.1	13.8	8,615
078B	MCG +11-19-016	56067	S?	15 48 08.9	+68 12 24	0.7 × 0.2		13.9	11.4	9,560
079A	NGC 6027A	56576	Sa pec	15 59 11.5	+20 45 15	0.7 × 0.5		13.9	12.5	4,296
079B	NGC 6027E	56579	S0?	15 59 14.7	+20 46 00	0.8 × 0.4		13.4	12.1	4,194
079C	NGC 6027B	56584	S0 pec	15 59 11.0	+20 45 41	0.4 × 0.3				4,117
079D	NGC 6027C	56578	SB(s)c? sp	15 59 12.1	+20 44 47	0.9 × 0.2				4,581
079E	NGC 6027D	56580	S?	15 59 13.2	+20 45 33	0.2 × 0.2		15.5	11.4	19,912
080A	CGCG 319-038	56588	S?	15 59 19.1	+65 13 58			14.5		8,980
080B		56590	S?	15 59 21.5	+65 13 22			15.4		9,601
080C		56572	S?	15 59 07.3	+65 14 01			15.1		9,567
081A	UGC 10319	57773	S?	16 18 14.3	+12 47 43	1.1 × 0.4				14,757
082A	NGC 6162	58238	S0	16 28 22.6	+32 50 57	0.9 × 0.7	60	13.6	13	11,221
082B	NGC 6163	58250	SB0?	16 28 28.1	+32 50 45	0.6 × 0.3		14.3	12.4	10,491
082C	NGC 6161	58235	S?	16 28 20.6	+32 48 38	0.9 × 0.3		14.6	13.1	10,139
082D				16 28 16.9	+32 48 48					
083A		58559	E0	16 35 36.4	+06 15 55	0.26 × 0.26				
083B		58565	E2	16 35 40.3	+06 15 44	0.3 × 0.2				
084A	CGCG 355-020	58877	S0?	16 44 23.0	+77 50 19			14.4		16,629
084B		58873		16 44 12.8	+77 51 32					
084C		58884		16 44 30.3	+77 49 48					
085A	CGCG 341-010	62476	E?	18 50 18.5	+73 21 05			14.4		11,079
085B		62477	E?	18 50 26.4	+73 20 42			14.9		12,046
086A	ESO 461-007	63753	S0?	19 52 08.7	−30 49 30	1.2 × 0.7	50	13.3	13	6,001
086B	MCG −05-47-003	63748	S0?	19 51 59.0	−30 48 57	0.5 × 0.4		13.8	12	6,023
086C	MCG −05-47-002	63752	S?	19 51 57.3	−30 51 23	0.4 × 0.3		14.9	12.7	5,356
086D	MCG −05-47-001	63749	S?	19 51 51.9	−30 48 30	0.3 × 0.2		14.7	11.7	5,743
087A	ESO 597-036	65415	S0° pec sp	20 48 14.9	−19 50 53	1.6 × 0.3	58	14.3	13.4	8,439
087B	MCG −03-53-003	65409	SA(r)0⁺ pec?	20 48 10.9	−19 51 20	0.7 × 0.6	100	14.4	13.3	8,717
087C	ESO 597-035	65412	S?	20 48 11.9	−19 49 54	0.7 × 0.2	94	15.1	12.9	
088A	NGC 6978	65631	Sb	20 52 35.4	−05 42 39	1.5 × 0.7	125	13.3	13.2	5,684
088B	NGC 6977	65625	SB(r)a pec:	20 52 29.8	−05 44 48	1.2 × 0.9		13.2	13.2	5,726

(continued)

Table 10.15. (continued)

HCG #	Other	PGC	Type	RA	Dec.	Size	PA	V Mag	Surf Brt	Radial Vel
088C	NGC 6976	65620	SAB(r)bc?	20 52 26.0	− 05 46 19	1.3 × 1.1		14	14.2	5,799
088D	MCG − 01-53-014	65612	S?	20 52 12.8	− 05 47 53	1.1 × 0.2		14.8	13.2	5,749
089A	MCG − 01-54-012	66570	S?	21 20 01.0	− 03 55 20	0.9 × 0.6		14.4	13.4	8,541
089B		66580	S?	21 20 19.2	− 03 53 46			15.1		8,676
089C		66575	S?	21 20 08.4	− 03 55 04			15.4		8,563
089C		66575	S?	21 20 08.4	− 03 55 04			15.4		8,563
090A	NGC 7172	67874	Sa pec sp	22 02 01.9	− 31 52 11	2.5 × 1.4	100	11.9	13.1	2,317
090B	NGC 7176	67883	E pec:	22 02 08.7	− 31 59 25	1.0 × 0.8		11.3	11.1	2,255
090C	NGC 7173	67878	E+ pec:	22 02 03.2	− 31 58 25	1.2 × 0.9	143	12	12.1	2,231
090D	NGC 7174	67881	Sab pec sp	22 02 06.7	− 31 59 30	2.3 × 1.2	88	13.3	14.3	2,509
091A	NGC 7214	68152	SB(s)bc pec:	22 09 07.6	− 27 48 36	2.2 × 1.4		12.7	13.8	6,529
091B	ESO 467-015	68164	S?	22 09 16.4	− 27 43 52	1.1 × 0.2	168	14.7	12.9	6,910
091C	ESO 467-013	68160	S?	22 09 14.0	− 27 46 57	0.9 × 0.8		14.3	13.8	7,020
091D	ESO 467-012b	68155	S0?	22 09 08.6	− 27 48 03	0.4 × 0.3		14.3	11.8	6,909
092A	NGC 7320	69270	SA(s)d	22 36 04.2	+ 33 56 52	2.2 × 1.1	132	12.6	13.5	464
092B	NGC 7318B	69263	SB(s)bc pec	22 35 58.5	+ 33 57 58	1.9 × 1.2		13.1	13.9	5,437
092C	NGC 7319	69269	SB(s)bc pec	22 36 03.6	+ 33 58 34	1.7 × 1.3		13.1	13.8	6,307
092D	NGC 7318	69260	E2 pec	22 35 56.8	+ 33 57 59	0.9 × 0.9		13.4	13	6,350
092E	NGC 7317	69256		22 35 52.0	+ 33 56 42	1.1 × 1.1		13.6	13.8	6,334
093A	NGC 7550	70830	SA0−	23 15 16.0	+ 18 57 41	1.4 × 1.2		12.2	12.6	4,751
093B	NGC 7549	70832	SB(s)cd pec	23 15 17.2	+ 19 02 30	2.8 × 0.7	8	13	13.7	4,336
093C	NGC 7547	70819	(R')SAB(s)0/a: pec	23 15 03.5	+ 18 58 23	1.1 × 0.5	106	13.7	12.8	4,470
093D	NGC 7553:	70842	E?	23 15 33.1	+ 19 02 51	0.7 × 0.6		14.7	13.6	4,768
093E	NGC 7558	70844	E?	23 15 38.3	+ 18 55 11	0.4 × 0.4		14.9	12.7	8,456
094A	NGC 7578B	70934	E1:	23 17 13.6	+ 18 42 29	1.0 × 1.0		14	14.1	11,702
094B	NGC 7578A	70933	S0° pec	23 17 12.0	+ 18 42 04	1.4 × 1.4		13.3	14	11,618
094C	PGC 70943	70943	S?	23 17 20.2	+ 18 44 05			15.2		11,769
094D		70936	S0?	23 17 15.3	+ 18 42 42			15.3		12,658
095A	NGC 7609	71076	Pec	23 19 30.0	+ 09 30 31	1.3 × 1.1		14.1	14.3	11,526
095B	MCG + 01-59-048	71080	S?	23 19 33.9	+ 09 29 43	0.7 × 0.2		15	12.5	11,126
095C		71077	S?	23 19 27.9	+ 09 29 40			15.2		11,278
096A	NGC 7674	71504	SA(r)bc pec	23 27 56.7	+ 08 46 43	1.1 × 1.0		13.2	13.2	8,405
096B	NGC 7675	71518	SAB(s)0−:	23 28 05.9	+ 08 46 06	0.7 × 0.4		14.9	13.3	8,257
096C	MCG + 01-59-081	71505	S?	23 27 58.7	+ 08 46 57	0.2 × 0.2				8,337
097A	IC5357	72408	SAB0 − pec:	23 47 22.9	− 02 18 03	0.9 × 0.5	150	12.9	13.5	6,622
097B	IC5359	72430	S?	23 47 37.9	− 02 19 02	1.1 × 0.2		14.7	12.8	6,587
097C	IC5356	72409	S?	23 47 23.8	− 02 21 05	0.8 × 0.4		14		5,642
097D	IC5351	72404	E?	23 47 18.9	− 02 18 48	0.5 × 0.4		13.6		5,886
098A	NGC 7783	72803	S0+?	23 54 10.2	+ 00 22 57	1.3 × 0.6	71	13	12.5	7,472
098B	NGC 7783B	72808	E?	23 54 12.3	+ 00 22 36	0.4 × 0.3	25	14	11.7	7,589
098C	MCG + 00-60-059	72810	E?	23 54 13.8	+ 00 21 24	0.2 × 0.2		15.3		7,791
098D	MCG + 00-60-060	72806	S0+?	23 54 10.9	+ 00 23 38	0.3 × 0.3				
099A	UGC 12897	54	Sab	00 00 37.9	+ 28 23 04	1.1 × 0.4	11	13.9	12.9	8,378
099C	MCG + 05-01-021	58	S?	00 00 44.1	+ 28 24 05	0.8 × 0.4		14.7	13.4	7,889
099B	UGC 12899	63	S0?	00 00 46.9	+ 28 24 07	1.0 × 0.9		13.7	13.4	8,543
100A	NGC 7803	101	S0/a	00 01 20.0	+ 13 06 41	1.0 × 0.6	85	13.1	12.4	5,011
100B	MCG + 02-01-012	108	S?	00 01 26.1	+ 13 06 47	0.8 × 0.6		14.3	13.2	4,902
100C	MCG + 02-01-009	89	S?	00 01 13.6	+ 13 08 38	0.9 × 0.5		14.9	13.7	5,110

Table 10.16. Blazars, quasars, and other exotic objects, courtesy of Steve Gottlieb/"Off the Deep End — 10 Challenging Observing Projects," and *Sky &Telescope* magazine

Object	R.A.	Dec.	Const	V mag.	Size	Notes
3C 66A	2 22.66	+43 02.1	And	13.5–15.6	Stellar	S&T, best seen Autumn to early Winter; near its brightest it appears as a 14th magnitude star
BW Tauri	4 33.19	+ 5 21.3	Tau	13.5–15.5	0.8" × 0.6'	S&T, best seen during late Autumn and Winter; appears as a fuzzy 14th mag. Star (the galaxy nucleus) surrounded by a low-brightness halo some 20" across
APM 08279+5255	08 31.7	+52 45	Lyn	16	Stellar	This mag 15.2 "star" is a superluminous QSO at a redshift of z = 3.9 (discovered in 1998), which implies a distance of roughly 12 billion light years!
Q0957+561	10 01.3	+55 54	UMa	16.5	Stellar	The famous gravitationally lensed twin quasars lies 15' NNW from NGC 3079. Individual components mag 16.5–16.7 separated by just 6"
HS 1046+8027	10 50.6	+80 12	Dra	15	Stellar	Relatively bright and nearby quasar (z = 0.12) discovered in 1999 and visible as a mag 15.0 "star"
Markarian 421	11 04.46	+38 12.5	UMa	12.0–14.4	0.8" × 0.6'	S&T, best seen Spring; appears similar to a mag- 12.7 star; small fuzzy halo glimpsed at 300×
Markarian 205	12 21.7	+75 19	Dra	14.5	Stellar	Markarian 205 is visible as a mag 14.5–15 "star" less than 1' S of NGC 4319. Arp redshift controversy with possible bridge to 4319
W Comae Berenices	12 21.53	+28 14.0	Com	13–17.5 (B mag.)	Stellar	S&T, best seen Spring; appears as a magnitude 14.7 (though it varies from 13 to 17.5) star
3C 273	12 29.1	+02 03	Vir	12.9	Stellar	Brightest quasar and relatively nearby at roughly two billion l.y. Appears as a mag 13 star in a triangle with a mag 13.5 and 15 star
SBS 1425+606	14 26.9	+60 26	UMa	15.8	Stellar	Very distant, mag 15.8 quasar (z = 3.165) at a distance of roughly ~11.6 billion light years
AP Librae	15 17.70	−24 22.3	Lib	14–16.7 (B mag.)	0.5" × 0.4'	S&T, best seen Spring; at high power (375×), it appeared as an extremely faint patch, just 6" across, punctuated by a dim stellar nucleus
PG 1634+706	16 34.5	+70 32	Dra	14.7	Stellar	Appears as a mag 14.7 "star" situated 1.6' SE of a brighter mag 13 star. This is very luminous, distant quasar with z = 1.337 (roughly 8.9 billion l.y.)
Markarian 501	16 53.87	+39 45.6	Her	13.5–14.0	1.1' × 0.9'	S&T, best seen during Spring; appears as a 14th magnitude knot, appears larger with averted vision. A "blinking" object that shows only a sharp stellar nucleus with direct vision, but the halo with averted vision
PG 1718+481	17 19.6	+48 04	Her	14.6	Stellar	Mag 14.5 quasar with a redshift of 1.08 and a light travel time of ~8 billion l.y. Forms the S vertex of a small triangle with 2 mag 13 stars
3C 371	18 06.84	+69 49.5	Dra	13.5–15.0	0.15'	S&T, best seen late Spring to early Autumn. Appears as a faint, quasi-stellar object that appeared softer than similar field stars. Higher powers show a compact 10" knot containing a 15th-mag. Stellar nucleus
HS 1946+7658	19 44.9	+77 06	Dra	15.9	Stellar	One of the luminous known quasars and visible as a mag 16 star at a distant of ~11.5 billion light years. A nearby mag 16 star confuses the identification
PHL 1811	21 55.0	−09 22	Cap	15.9	Stellar	Recently discovered quasar (2001) is the second-brightest known at B = 13.9, and with a redshift of z = .192 is 20 % more distant than 3C 273
BL Lacertae	22 02.72	+42 16.7	Lac	12.7–16.0		S&T, best seen late Summer to Autumn; sometimes appears slightly fuzzy at high powers, with a tiny 2" to 3" envelope

Notes: "S&T" means that the source of this line of data is *Sky and Telescope* magazine, April 2010 edition, pp. 71–74; these data and descriptions used by permission from *Sky and Telescope* magazine editors. The object descriptions (paraphrased from these pages by Steve Gottlieb as he saw them with his 18-in. telescope and various magnifications)

Detailed descriptions and data source is Steve Gottlieb

Quasars and Exotic Distant Objects

Finally we look at some of the most extreme, exotic objects in the universe. Table 10.16 provides just a sampling of these, with data obtained from both Sky and Telescope and Steve Gottlieb. All of the descriptions were provided by Mr. Gottlieb. As you are viewing these objects, consider the immense power that makes these unusual objects shine, and the fact that we are able to enjoy these deadly splendors at a safe distance so as not to suffer the ill effects of their radiations. Consider how awesome these objects are; we are witnessing the "flame out" of material being consumed by a black hole that carries millions, if not billions, of Sun's worth of mass.

Now that you have decided to pursue some of these observing lists, you may want to keep a record of where you've been and what you have accomplished. We will go into detail next on how you can do just that, and how you can make scientifically useful observations (or make scientific use of your observing records) of faint objects in the next two chapters.

Chapter 11

Recording Your Observations and Other Tips to Help You Stick with the Program

Introduction

Having made your observations, how will you remember them? The best way to do so is to record the observation by drawing (or imaging) the objects and including a detailed description of the objects along with sky conditions, date, time, instrument, and magnification used. With the widespread availability of digital cameras for astronomical purposes, along with various filters, both broad- and narrowband, many readers may opt to take images of faint objects. Imaging also brings out still fainter objects that are not available visually. We will leave the details on how to do imaging to another source/author, but know that imaging faint objects will require long exposures to capture these objects in detail.

We will focus primarily on the visual observer in this chapter, but several parts of this chapter, especially the documentation part, applies to images. For instance, it is important to record the details of your setup, to include telescope and camera type, the make and model of the CCD chip, the computer programs you use to capture and process your images, and the processing steps you used to produce the finished product. It is useful to record exposure time(s) and the prevailing weather conditions and sky transparency to get a sense of what an appropriate exposure time for a given object under a given set of conditions. If you are doing narrowband imaging (imaging deep sky objects with filters of S II, O III, H-alpha, or others) it is important to know what the appropriate exposure of each should be.

In fact, narrowband imaging, if done properly, may be scientifically useful. It would be interesting to have a catalog of images of various nebulae taken under similar conditions with the same instrument setup, showing the distribution of the various chemical elements within each object over a large number of objects and to look for patterns and trends therein. We provide this as a suggestion for the narrow band-imagers out there (Fig. 11.1).

B. Cudnik, *Faint Objects and How to Observe Them*, Astronomers' Observing Guides,
DOI 10.1007/978-1-4419-6757-2_11, © Springer Science+Business Media New York 2013

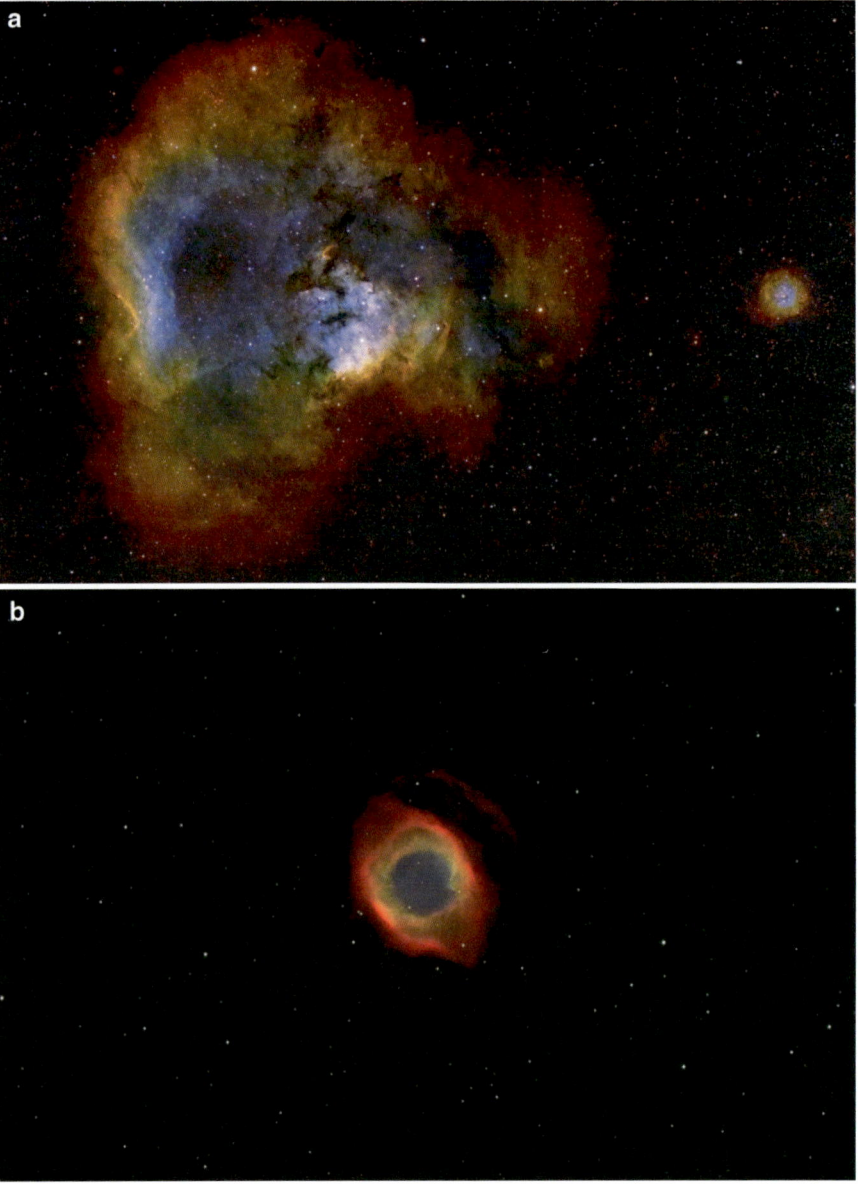

Fig. 11.1. Two examples of narrowband images of nebulae (**a**) emission, Sharpless 2–171; and (**b**) planetary, the Helix Nebula NGC 7293). The different colors represent different elements composing the objects: e.g., *blue* is OIII, *red* is SII, and H-alpha is *green* (Images courtesy of Don Taylor).

Specifically here is what you should keep in mind as you record your observations:

- The details of what you record at the eyepiece will depend on the objective of the observer (you). Is this for personal use only or are you working toward an award of some kind?

- Record as much information as possible and include a sketch, so as to provide a record of your experience.
- If you are working toward an observing club award, be sure to include all required information.
- The more detail/information (including drawings) is recorded, the easier it will be to revisit (and relive) the observation in the future.
- Stick to your plan to do the list in stages, grouping the list into manageable chunks. Also, don't simply rush through the list in order to complete a checklist; rather, take time to appreciate each object, its physical nature and its subtle beauty

Logging What You See

You should keep a log of what you see. This author been doing so since 1980 with a wide variety of astronomical subjects: solar motion, lunar motion, telescope views of all kinds of objects, aurorae, meteors, the weather conditions, and much, much more. In fact, each semester I teach introductory college level astronomy lab, I have the students look for various astronomical objects and log what they see. I still log most of my astronomical observations to this day.

Make the observing log something that you will get meaning out of and enjoy. You may look at others' logs for ideas on how to set it up, but the work should be your own, with your own observations. Make a sketch to represent your observations. There are generally two types of framing to contain your sketches on the paper: a circular frame that represents the perimeter of the eyepiece field of view, though you will rarely exactly map one-to-one the actual field of view and your sketch, as you are likely to tend to "zoom in" a bit, covering a smaller area of sky in your drawing than what is actually viewed. This arises from another tendency most visual observers have of enlarging their target object.

You can also have a rectangular or square frame that includes only the area immediately around the location of the object and not necessarily the whole FOV. The former has the advantage of being more realistic; the latter enables more observations to be fitted to the page. The drawings presented in this book use both types of framings. The basics of astro-sketching can be found in other books, which describe it in detail. Here, we are assuming an average experience level of moderate (with some advanced and some beginner) and some experience at logging visual observations.

There are two basic ways to record your observation: at the eyepiece by writing notes (or writing notes IMMEDIATELY AFTER observing to prevent your impressions from being distorted by imagination) or speaking your description into a mini tape recorder. Though many prefer writing notes, the tape method does allow you to describe the object without having to take your eye off the eyepiece. Plus it prevents the need for turning on a red flashlight to write; the red flashlight may reduce (slightly) your dark adaptation so as to render objects that would be barely visible to the fully dark adapted eye invisible. If you chose to write, do so immediately after and not during an observing session.

Since you are "copying" what you see at the eyepiece when sketching, this necessitates looking back and forth from the eyepiece to your sketch pad (with the red light on…). Here are some tips to help with this, especially with faint comets:

- Take time to study the object fully before committing pencil to paper [97]. Once you are satisfied with the time spent studying, then begin the process.
- First render the object at the center of your frame-on paper.
- Next place the stars around the object as appropriate. The object may become invisible because of your eye's exposure to red light, but memorize its location among field stars so you can place them accordingly.
- Write down the date; time (either starting and ending, one or the other, or mid-point); location and altitude of object; seeing (1–10, 1 being the worst, 10 being the best possible, or "perfect"); transparency (1–6 on the Bortle scale, 1 being the worst, 6 being very transparent); telescope type; aperture; f/ratio, magnification/eyepiece, filter(s) used; and technique(s) used. Record your impressions in as detailed a form as possible.
- Describe the object (and possibly the surroundings). Use graphic expressions but DON'T exaggerate!
- Include field orientation in your sketch. You can tap the eyepiece diagonal in an N-S or an E-W direction and note how the field moves. Or you can practice the drift method. If you have two stars, a bright and a faint star next to each other oriented roughly E-W, turn off the clock drive (if you have one) and let the field drift. The star that leads the pair is the westernmost one and all the objects will move in a west direction. If you are using a Schmidt-Cassegrain telescope, north is up if direction of motion is left, and vice versa. If using a Dobsonian reflector north is usually down if the field motion is left (as is usually the case). When unsure whether something is north or south, nudge the tube toward the NCP and the site where stars enter the FOV. Be sure to move the tube in the direction perpendicular to the drift.

You can use the following questions to help you write your descriptions [98]:

- For galaxies, is the core bright, diffuse, or stellar?
- Can you see structures/textures in the galaxy or nebula you are viewing, or does it appear smooth/uniform (e.g., spiral arms, dust bands, and knots)?
- Is the edge sharp or diffuse (or both)?
- Are there stars superimposed on the object?
- Are there any other deep sky objects in the field of view?

In most cases you may not be able to answer all but a few of the questions, many of which are geared toward galaxy observing, which most of your objects will likely be, but they can apply to non-galactic objects as well.

Whatever you do, be as truthful and accurate as possible; don't make up stuff!

Additional Information to Help Enhance Your Observing

You can estimate field size by "measuring" it with a wide binary star of known separation, or by using the Moon (assuming you know the angular size it is subtending that night, but you can also use its average apparent diameter); you then estimate the fraction of size of the FOV the object appears. A word of advice if you use the Moon for your "measuring rod": do so on a night you are not planning to view faint objects, as the Moon will certainly ruin your night vision. Or you can use the Moon during twilight to gauge the fields of views of the eyepieces you will be using (you will need to select them ahead of time to do so). Be sure to take notes on these exercises, to prevent memory lapse. Be sure, if possible, to look up the Moon's apparent diameter for the day you do your measurements.

A more accurate method to measure your eyepiece view field diameter is to time how long a star will take to drift from one side of the field, through the center (or as close to the center as you can get) to the opposite side. Pick a star near the celestial equator to make the calculations as simple as possible. The standard formula for calculating apparent size ($diam.$) is given as follows: $diam. = 0.25t \times \cos(dec)$ with t being the drift time of a star from one end of the field to the other in seconds and dec being the declination of the star used. If you use a star near the celestial equator, $\cos(dec)$ become 1 and the formula is simply $diam. = 0.25t$.

Alternately you can do all this ahead of time before you even get to "the field." Websites such as http://www.csgnetwork.com/telefov.html have a built-in calculator to measure the eyepiece field of view. If you have a planetarium program that shows stars down to the limiting magnitude that you can see in your eyepiece, match up the view, and use separations between field stars as a gauge of size. *TheSky 6* allows you to match the eyepiece field and use patterns of stars within that field as little rulers to gauge the size of something.

Once you get a comet in the FOV, go to the computer and zoom in on the simulated FOV until you are able to match that field with the eyepiece view. Find a small pattern or two of stars in the telescope's eyepiece FOV, and use the matching simulated FOV on *TheSky 6* to measure the separation between pairs of stars. Then use the pair with known separation as a 'ruler' to gauge the size of the object.

You can make a table of characteristics of eyepiece views with each eyepiece and telescope (Be sure to do all your telescope/eyepiece combinations.). You might use one table for 'scope A, a second table for 'scope B, and so on to include all of the 'scopes you use regularly. If you do more than one type of astronomy (e.g., bright objects, variable stars, lunar and planetary, solar, etc.) and you have telescopes dedicated to brighter objects, include these as well. You can spend a cloudy night setting up and filling in your eyepiece tables for each of your 'scopes. Include each eyepiece that you use with each 'scope (and you may use the same eyepiece for several 'scopes), labeling the eyepiece by brand and focal length in mm. Use the above website and websites like it to calculate values for field of view, apparent field of view, etc.

At the top of the page containing the table, include your telescope's make, model, aperture, f-ratio, focal length, and primary use(s). For each eyepiece that you will use with the telescope, include as many of the following as you wish:

- Eyepiece brand, type, and focal length in mm.
- The size of the eyepiece: 0.965-in. (rarely used for deep astronomy nowadays), 1.25-in. or 2-in.

- Magnification the eyepiece provides (to get this, divide the focal length of the telescope, which is aperture × f-ratio × 25.4 mm/in., unless your telescope aperture is given in mm).
- Actual field of view in arc minutes.
- Apparent field of view in degrees.
- Filters that are compatible with the eyepiece (that is, the threads of the filter match the eyepiece so that you can screw the filter in the back of the eyepiece).
- Limiting magnitude viewable with the eyepiece under "normal" conditions at your dark sky site (and in your back yard if you do astronomy from a light-polluted location regularly).
- Primary use(s) of the eyepiece (e.g., thin galaxies, planetary nebulae, etc.).
- Any other comments you find helpful.

In terms of keeping and using records of your observing sessions, there are a number of things that you can do with your drawings once you complete a session. Besides having a record of "where you have been" in your exploration of the universe, you may want to do statistical studies of your results [99]. You may wish to submit your drawings to magazines such as *Sky & Telescope* or *Astronomy*, or publish them on a website or use them to write a guide for future observers. There is even some scientific work that you can do with only your 'scope, your eyes, the appropriate charts, and some time and dedication.

Become the local "expert" on the objects you have observed, either as a class of objects or on a handful of individual objects. Lots of information is available online; you can learn about your chosen topic, and as you observe the objects a second (or third or an additional) time or continue on your quest to observe as much of the universe as possible, you will more deeply appreciate your time behind the eyepiece. The edge of the universe (not just the sky!) is your limit on what you can accomplish with your newfound program of observing faint objects.

In addition to your own observations, you may want to help in the actual scientific research that seeks to know more about the objects that you have painstakingly sought out in your observing projects. We will go into more detail in the next chapter on how you can contribute to the science of astronomy.

Some Personal Observations

First, however, some examples of the author's own observations of faint objects will be presented, and will include descriptions and drawings. Perhaps you can follow my examples, or take these examples and adapt them to your own projects. All of these were made visually from the Dark Site Observatory during the spring and summer of 2011, with a 14-in. (35.5-cm) f/11 Cassegrain at 98× unless otherwise indicated (Fig. 11.2).

Here are the objects observed and their descriptions:

- *NGC 2537*. Easily seen as a fairly large patch.
- *IC 2233*. Seen only with averted vision; looks like a needle.

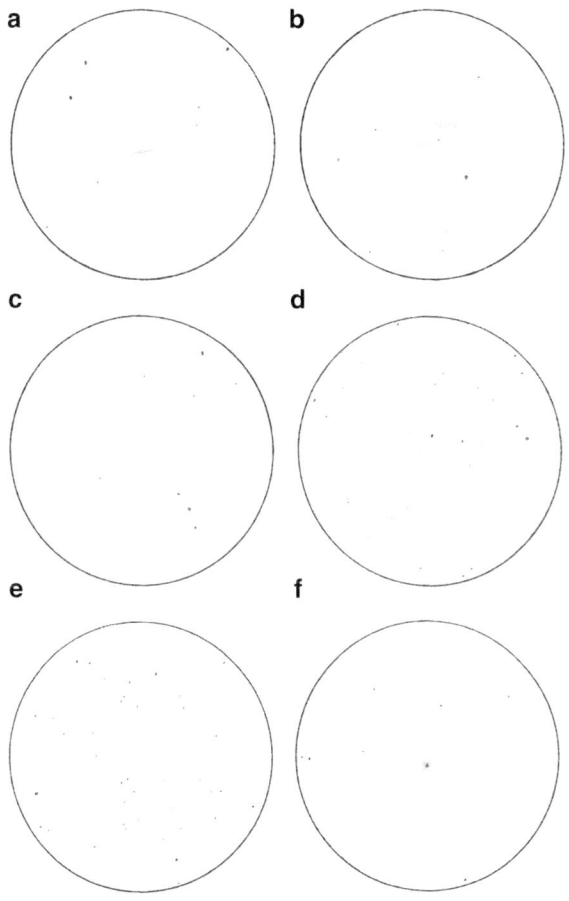

Fig. 11.2. (**a–f**) Author's drawings of faint objects: (**a**) the thin galaxy UGC 7321, (**b**) Jones-Emberson (Jn-Er) 1, (**c**) the nearly edge-on galaxy NGC 5529, (**d**) the faint open cluster NGC 7044, (**e**) the faint globular cluster NGC 6749, and (**f**) the faint spiral galaxy NGC 5989.

- *NGC 2437A.* Too dim, not seen.
- *NGC 2424.* Appears as faint or very faint needle.
- *Jones-Emberson 1(PK164 + 3.1).* A planetary nebula, which appears large and quite faint; the west northwest half appears brighter than the eastern half; I used an OIII filter, averted vision, and telescope jiggle to make this one at least faintly visible.
- *UGC 7321.* Very difficult with a C14, as the object was 40° up under Bortle class 4 skies. Barely visible line segment with averted vision, needed (and used) a photograph to verify the object's location.
- *NGC 5529.* The brighter of the faint objects, easily seen but rather faint, widens gradually toward the center. A hint of mottling noticed at high magnifications.
- *NGC 5545/5554.* Appears as two hazy spots in contact with each other. One appears somewhat brighter than the other but, although faint, can be seen relatively easily. Both appear more round than flat with only modest condensation.

- *Palomar 8.* Appears as a small roundish mottled gray haze in a rich star field (stars of various brightnesses as opposed to all being about the same brightness). This one was easy to spot through the C14 under dark skies and at 98× magnification.
- *Tonantzintla 2.* Very difficult object to observe due to its low elevation (between 12 and 20°, the maximum for my location), it appears as a small, very faint, very vague haze mixed with and surrounded by an oval of faint stars and an 11th magnitude beacon at the south southeast end. This object was observed through the C14 at 98× and 244.5×.
- *Djorgovski 2.* It lies west (about 20 arc minutes WNW, actually) of the beautiful pair B86 (dark nebula) and NGC 6520 (open cluster) in a starcloud field. It appears as a faint haze with averted vision mixed in a triangular configuration of faint stars. Observed through the C14 telescope at 244.5×.
- *Sharpless 2-12.* With the 12.5-in. f/7 reflector and a 40 mm eyepiece with OIII filter, I swept the sky west of the beautiful open cluster M6, starting with the cluster itself. I centered on the fairly bright "central star" of the complex which itself is center of the NGC 6383 open cluster. I used averted vision and was able to pick up the entire eastern half, visible as a vague, half-moon shape with the "terminator" running north–south across 6383. This seems to be the brightest half. This object was 26° high in the south at the time of observation.
- *Sharpless 12-13.* This object was easy to locate in terms of its position, with three medium bright stars lined up south southeast to north northwest. The nebula was seen as a vague glow between and including the southernmost two of the three. It appeared quite faint but was still definitely visible. The object was about 27° high in the south at the time of observation.
- *Sharpless 2-71.* Even with the OIII filter this object looks little more than an elongated smudge with some brightening toward the center. The observation was made at 98× through the C14.
- *Vyssotksy 1-4 (PK 27-3.2).* This is illustrated in a previous chapter (Fig. 6.1) that illustrates the effect of using an OIII filter. This is a case where the object may not be obvious at first. It looks like two faint stars close together without the filter, but the southern one does look a bit "bigger" than the northern one at 244.5× with the C14. The difference became obvious with the OIII filter as the nebula shone through brightly, but the surrounding stars were dimmed considerably.
- *NGC 6749.* Appears rather indistinct, but it does appear to stand out a bit from the background. Seems a fairly large loose collection of mostly faint stars mixed with nebulosity. The position and size were confirmed with an image, making the identification of this object certain. The skies were clear with a bit of haze and a limiting magnitude of 6.2.
- *NGC 5989.* This appears quite faint, indistinct, but I am looking at it as it stands 30° high in the western sky. I did need the chart of the region by the program *TheSky 6* to pick out this object, which was observed at 244.5× with the C14.

Staying with the Program

Before we jump into the scientific side of observational astronomy (Chap. 12) here's some helpful information on how you can stick with your faint objects (or other type of) observing program. Following these practices, as they apply to you, will maximize not only the chances of you completing observing projects but also enjoying the observations of faint objects (or whichever area of astronomy you become interested in) for many years.

As you embark on your program, don't try to do too much too fast. In so doing, you run the risk of burnout, which then means that the hobby is no longer enjoyable. If the experience ceases to be enjoyable, stop for the moment and take a short break from the action. The break can range from a few minutes in a warm room away from the eyepiece or a few hours' nap before resuming observations, or for a more extended time. Doing this can re-establish your appreciation for the night sky. Related to this is the suggestion that you do not make your lists too long or your expectations too high as you begin your program. If you need to, start with the easier objects first, and then work your way to the more challenging objects. Don't start with the faintest objects, or you may get discouraged.

In fact, if you are just starting out, focus on the brighter objects at first, then work your way to the fainter, more challenging objects. Mix it up: don't just do faint objects, but look at bright objects and Solar System objects (and the Moon) as well, but be mindful of preserving the best of your dark vision for really faint and bright objects. Do some binocular/wide field observing as well as focused telescopic deep sky observing. There are many deep sky objects such as the North America Nebula that actually look better through a good pair of binoculars than through a typical telescope.

Take time to "smell the roses" or "absorb the photons"; that is, use time to study your object and appreciate its beauty and nature. And to help you better appreciate these things, join up with a network or a local astronomy club to find like-minded individuals to collaborate, share ideas and observe with. They can encourage you or provide guidance to help you get even more out of your observing experience.

Subscribe to magazines such as *Sky & Telescope*, which offers something new to observe each month. And you can use these to help you learn more astronomy, especially about the objects that you are observing. You will find that the more you know about the inner workings of an object, the more you appreciate an otherwise unremarkable patch of haze that you can barely see.

Incorporate an award or club with your observing program. Also, have a goal you are working towards. This motivates you to keep going to completion. But as you are going, don't let a string of cloudy nights discourage you, and if you need to take a break, do it! Sometimes the best thing our hobby needs is a break to rekindle the passion and rest the body and soul.

Conclusions

You can get a lot more out of observing faint objects (and any other type of visual astronomy) if you record what you see. By keeping a record of your "adventures," you document what you have seen and where you have been and have a means of sharing your observations, your experience with others, and even comparing notes. On cloudy nights, you can review your observations and perhaps plan new adventures in the realms of "extreme astronomy" or otherwise. In this chapter we have provided some methods, approaches, and a few examples of to help get you started.

Not only can you record your experience with observing faint objects by keeping a log, you can also document your progress in terms of your ability to see faint objects. It also would be interesting to have a series of entries of observations made with different instruments of the same object. That way you can see, and record, for yourself the effects of aperture on the amount of detail you see.

In addition to this, making a record of your observing experiences will likely help you stick with the program. We have provided several hints at what you can do to keep your passion for astronomy alive and active, and sometimes that involves taking a break from the action. Whatever you do, do your best to not let the spark of interest fade. Even if you move to a different area of astronomy, away from faint objects, stay the course. There's a big universe to explore!

Recording Your Observations and Other Tips

Chapter 12

Citizen Science Activities and Searching for Supernovae

Introduction

There are plenty of opportunities for everyday citizens to contribute to the science of astronomy while searching for faint objects. The observing lists and information presented up to this point outline how you can get started and keep with the program of observing faint objects from a more recreational perspective. Although there is indeed much value in pursuing objects in that capacity, one can also go farther in one's pursuits if one wants to join the ranks of citizen scientists all over the world who are contributing in real and valuable ways to the science while pursuing their favorite hobby. If this describes you, then the following sections will be useful for you to expand your astronomical significance.

Glossary in this book provides information on books and resources on the scientific side of astronomy. The following will give an overview of what opportunities are out there. I also share my own experience in this area. Although I have had some experience with professional research (I have published several refereed papers in the areas of solar physics, lunar meteoroid impacts, and Comet Hale-Bopp) I prefer the comparatively relaxed pace of being a citizen scientist. There is much value in the many forms of professional research ongoing today, and I have several such projects pending, but there is an increasing need for citizens to help with the increasing load of data produced.

Projects Worth Pursuing

Variable Stars

My primary activity is making magnitude estimates of variable stars, to include the occasional (sometimes three at a time in the same area of sky as was the case in 2011) supernova. There is still the need for visual magnitude estimates of variable stars, to maintain continuity with historic data and continue to corroborate with

B. Cudnik, *Faint Objects and How to Observe Them*, Astronomers' Observing Guides, DOI 10.1007/978-1-4419-6757-2_12, © Springer Science+Business Media New York 2013

electronic data. I observe primarily stars that are in need of more data and objects in support of various observing campaigns. Also, many of the surveys that measure the magnitudes of stars thousands of times per night can only work with stars of the 12th magnitude and fainter, as brighter stars will saturate the detector. Also, some of the surveys avoid the Milky Way entirely due to the large numbers of bright stars, hence the need for visual and small aperture photometry that amateur astronomers can provide. There is also still a great need to observe their program of stars in need of more data through their minima when they can glow at 14th, 15th or 16th magnitude. These objects in need of more observation certainly qualify for faint objects.

For a primer of how to observe variables, www.aavso.org (the American Association of Variable Star Observers, AAVSO) has a nice brief tutorial. For more information, books such as *Variable Stars and How to Observe Them* (Springer 2010) are available to provide an in depth introduction to the types of variable stars that are out there, as well as tools and techniques of observing them. Note: If you are new to variable star observing consult these resources first and gain some experience observing the suggested stars they provide for beginning observers. The stars alluded to above are for more experienced variable star observers.

The AAVSO is looking for people who are able to visually observe and make magnitude estimates of faint variable stars or stars that are faint near their minima. It takes experienced observers to make quality magnitude estimates of faint variables, and AAVSO needs more of these types of observers. Stars in need of more visual data are usually poorly observed near their minima due to faintness, and observers are needed to help fill in some of these gaps. There are other stars on their list that stay faint all the time. They also need observations of novae and supernovae past their peak, when they are faint (and fading, usually several months after their peak and after most observers' interests have peaked), and monitoring of faint dwarf novae to characterize their periodicity (or lack of...). More information, including real-time alerts, lists of stars needing more observational coverage, and predicted dates of maxima and minima are available at the AAVSO website, www.aavso.org.

The really good part about variable star observing is that it combines several aspects of astronomy: faint objects (for the aforementioned stars), variable star observing, effective logging/record keeping and citizen science activity (definitely a motivation to share your observations). So if you are ready to embark in this new direction, go for it! Here is something else you can work on while you are at it (next section).

Supernova Searches

Supernovae (Fig. 12.1) happen every 50 years or so in a given galaxy [100]. If you watch S0- and SB0-type (lenticular) galaxies, you have a somewhat better-than-average chance of spotting one of these with a supernova. The Reverend Robert Evans of New South Wales, Australia, discovered 14 supernovae from 1981 to 1988, and also co-discovered 3 supernovae, most of these made with a 16-in. (40.6-cm) telescope. He did all of this visually and memorized hundreds of star fields so as to be able to quickly pick up any interloper (or supernovae or transient) in the field. In 2008, then 14-year old Caroline Moore of Warwick, New York, discovered SN 2008 ha, which turned out to be surprisingly faint—1,000 times fainter than the

Fig. 12.1. SN 2011fe (*arrowed*) in M101 on August 25, 2011 (Image courtesy of Thunderf00t, used by permission under creative commons license, attribution).

typical supernova. As one can imagine, this has important theoretical implications for how supernovae work. Moore became the youngest person ever to discover a supernova [102].

Following are some general guidelines to begin and sustain a viable supernova search program. First, find a resource (an example is Supernova Search Charts and Handbook, Thompson, G.D. and Bryan, J. D., © Cambridge Univ. Press, 1989) or two that will provide images to compare with eyepiece fields of view to look for supernova candidates. Another useful resource, written and compiled by Juhani Salmi, is entitled *Check a Possible Supernova*, which consists of two sets of bound index cards that consist of an image of a galaxy per card. The purpose of these is to enable an observer to compare the field of view of the galaxy that one sees through the eyepiece with the image (black on white negative, which makes it much easier to use at the eyepiece) of that same galaxy. If you do not have or are unable to get these resources, you can look for other references, such as images of galaxies to use in making the above-described comparison, images that include black stars and galaxies on white backgrounds, with similar limiting magnitude as your instrument and sky setup.

If you find a candidate supernova, check it multiple times over a period of several hours, if possible, to ensure it is not an asteroid that happens to be crossing the galaxy as seen from Earth at the time of your observation. A sketch of the galaxy and all the visible stars, with the suspect highlighted, and all the objects carefully placed on paper, will help verify the nature of the object. If it is an asteroid, try to identify the asteroid (in case it is a new discovery, which the Minor Planet Center, http://minorplanetcenter.net/iau/mpc.html, handles); if it is a supernova, report it at once to the AAVSO, including as much detailed information, such as distance of supernova from the nucleus in terms of E-W and N-S (and line of sight) distances in arc seconds, the rough estimate of the apparent magnitude of the object and any changes in the brightness, and other pertinent information.

If you plan to search for supernovae, it is a very good idea to be familiar with how bright stars of representative magnitudes appear in your instrument at a given magnification, to include typical limiting magnitude. Although this may vary, if you use the same site under similarly dark conditions, this becomes easier to do and could potentially help to provide more accurate initial brightness information on your supernova candidate. Patrolling a set of galaxies on a regular basis, bright ones usually, is an excellent way to search for supernovae. As has already been mentioned, it is an excellent idea to be familiar with the territory, and as you revisit your galaxies again and again, the star patterns associated with each galaxy become familiar as constellations in the nighttime sky, which could make spotting a supernova candidate a quick exercise, should one appear. This project is especially useful if integrated into existing activities; so, while enjoying the view of your favorite galaxies, be on watch for "extra stars" that could be a supernova in progress.

For existing supernovae, one needs to follow existing objects as long and as faint as one possibly can, which contributes to the light curve (an example of which is shown in Fig. 12.2 for SN 2011FE). Another related project that one can pursue is to search for supernovae that have not yet been found. Although imaging does tend to snap up new discoveries, there is still room for regular visual inspections of the brighter (and even the not-so-brighter) galaxies for supernovae eruptions. One may start with S0 and SB0 (lenticular) type galaxies, which tend to host supernovae more frequently (according to North [101]), then expand to ordinary spiral and barred spiral galaxies.

Classical novae erupt from time to time and will also need to be followed, especially as they fade and become quite faint. This is especially true for the beginning phases of their eruption, but the entire nova outburst, from start to finish, should be followed closely. Dwarf novae and eruptive variables, especially if they are not regular (the case of BW Sculptoris in the fall of 2011 is one such example) are particularly in need of observational coverage. Most precursors to these are faint and need to be observed regularly, especially if you can see the object (preferred), and this is, again, another example of a faint object to observe with potential scientific benefits attached.

Citizen Science Activities

In recent years, there has been a greater and greater need for ordinary, yet interested, members of the general public to help out with some aspect of scientific research [103]. "Citizen Science" refers to a practice that dates back to Sir Isaac Newton (and even before) but has received more widespread attention only in the last few decades or so. Several factors have come into play to make this happen,

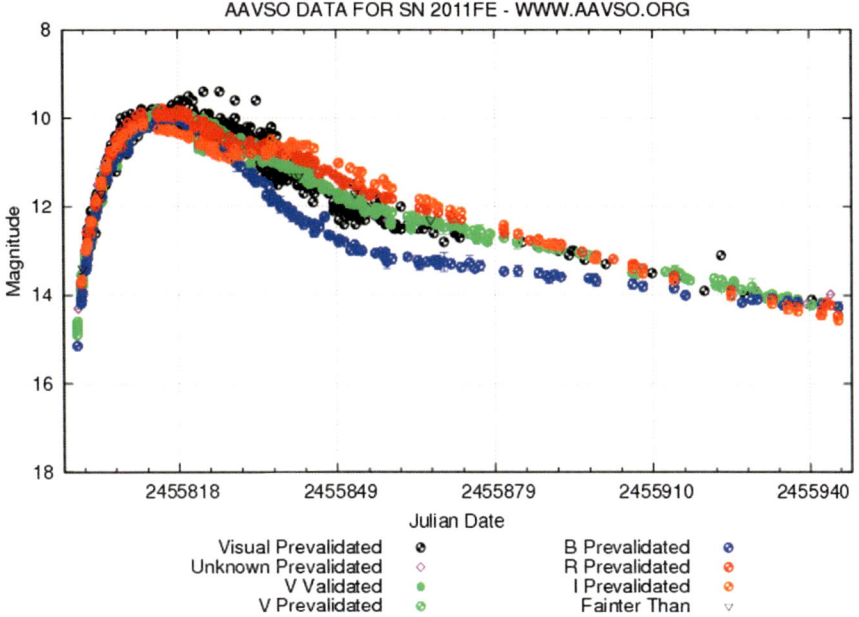

Fig. 12.2. Light curve of SN 2011fe generated by the AAVSO Light Curve Generator (LCG) (Courtesy of AAVSO. The author contributed a number of data points to this light curve which starts 24 August 24, 2011, and continued to 20 January 20, 2012).

including the widespread availability of CCD and photometric technology (once reserved for the professionals at institutions) to the public, the explosion of astronomical data from various surveys needing to be analyzed, and limits on how much a professional astronomer can do. Recent changes in practice have increased the scientific soundness of public-contributed data as well as measurable goals to provide guidance and a gauge of progress. There are only so many trained professionals, and funding can be a limiting factor.

Professionals have a variety of constraints that limit the amount of data that they can collect and the time they can spend collecting data, but a well-organized network of adequately equipped amateurs can overcome these hurdles. Citizen science has been formally defined as "the systematic collection and analysis of data; development of technology; testing of natural phenomena; and the dissemination of these activities by researchers on a primarily avocational basis [103]." Different forms that citizen science has taken in the areas of astronomy and space science include the following:

- Collection of data to be analyzed by professional researchers. Examples include variable star magnitude estimates and the timings of lunar and asteroid occultations.

- Assist in the analysis of data collected by professionals through distributed computing (participants' computers in their idle time can be set up to analyze chunks of data from SETI, Einstein@home, etc.) or in person using a computer. Examples of include SETI (Search for Extra-Terrestrial Intelligence) Live, Galaxy Zoo, and other Zooniverse projects.

- Volunteer to assist at a research center.

- Compete in an organized competition that would benefit technologically. An example is NASA's International Space Apps Challenge.
- Build and operate their own instruments to collect data for their own experiments or for part of a larger project or collaboration. Some examples include the Solar Ionospheric Disturbance (SID) program of the AAVSO Solar division and amateur radio astronomy activities.
- There are also opportunities for citizen scientists to do sub-orbital experiments and develop payloads as part of a competition; see http://www.citizensinspace.org/ for more information.

As we have seen, there is a wide variety of citizen science projects available, and this is not limited to astronomy, either. People interested in contributing to other branches of science can check out the website http://buhlplanetarium2.tripod.com/FAQ/citizenscience.html, which has a list of many such projects in areas such as computer/electronic, astronomy/space science, bird watching, insect watching, weather/environment, animal observing, and flowers/plants. You can find astronomy projects by going to http://buhlplanetarium2.tripod.com/FAQ/citizenscience.html#astro. This has links to many of the organizations involved in citizen science activities. The Citizen Sky website, http://www.citizensky.org/, focuses exclusively on astronomical projects, with the main focus being the recent minimum of the enigmatic eclipsing binary star epsilon Aurigae.

Again, websites tend to change. If one of the sites provided here is no longer active, you can look for citizen science projects (assuming they still exist) by doing a web search for "Citizen Science."

Here are a few more projects that one can consider:

- Gauging the quality of an observing site in terms of transparency over time. You may wish to do this with your chosen dark site to find any long-term trends of whether your site is staying dark or if it is growing brighter over time (as is the case in most places).
- Secrets to success in challenging programs… a few are given here to get started. You may wish to perfect this and add to your own—and organizing and publishing your results in a magazine such as *Sky & Telescope* or *Astronomy*. You may also wish to share your experience with a local astronomy club.
- What physical factors go into determining the visibility of an object, and how does this change with higher magnification?

Serious searches for supernovae should be done as often as possible, and each session is well documented to maximize the scientific worth of any discovery that takes place. If a supernova is seen or suspected, contact the International Astronomical Union's Central Bureau for Astronomical Telegrams—go to their website (http://www.cbat.eps.harvard.edu/HowToReportDiscovery.html) and follow the instructions on that page to make your report. By following these instructions, you will ensure that your discovery will reach the largest number of people as soon as possible, and will also make it easier for other observers to confirm your discovery. If you think you found a new variable star of another type, follow the instructions provided on the AAVSO website to ensure it is truly a variable star and not a solar system interloper. The website is http://www.aavso.org/how-report-new-variable-star-discoveries.

Conclusions

One more way to make observational astronomy interesting is to make use of your observations, whether it is by a simple statistical analysis of your work or sharing your observations and techniques (after practice…) with your observing friends, local astronomy club, or *Astronomy* magazine, or making scientifically valuable observations and sharing your work with others.

As discussed above, there are many opportunities to become a citizen scientist and enjoy faint object observations. From classifying faint galaxies to observing faint variables and supernovae, there is a wealth of opportunities to expand your faint objects experience and to make a lasting contribution to the scientific side of astronomy. As hinted at above, there is a certain connection one gets when looking directly at something as opposed to imaging it (which does have tremendous value as well). There is still a place for visual astronomy of faint objects even in the twenty-first century.

A Few Final Thoughts

Throughout these pages, we have explored the universe of faint objects and outlined in fair detail what to see and how to go about seeing these. Here are some final tips for observing faint objects:

- Pre dark adapt. Sunglasses even after twilight ends preserves best sensory perception.
- Use cloth black hood when viewing through eyepiece.
- Use paper finder charts and a very (extremely) dim red light for very short time periods.
- If you use a display screen at the 'scope, use a very deep red filter over it and look at it for very short times (don't stare at it).
- Don't observe near others that have any lights in their area.
- Take dietary supplements. Some claim that bilberry reduces floaters in the eyes.
- Find the best averted vision spot in your eyeball.
- The obvious is averted vision. Scan around the known location of the object until you find and learn where your "averted sweet spot" is.
- Use an eye patch over your non-observing eye to keep face muscles relaxed… and keep both eyes open.
- A little 'scope tap will many times bring an object into view, just a little nudge at the top of the UTA to get the stars moving around a bit.
- Experience, experience, experience, observe, observe, observe, etc.

In this book we described the physical nature of deep sky objects in general, that is, objects that lie beyond the local Solar System. These were the focus of this work. "Faint objects" are a subclass of the various forms of deep sky (or "deep space") objects. In fact, "faint objects" make up the vast majority of deep space objects that we can see from Earth or from low Earth orbit. If one takes a tally of, say, galaxies, one would find that most are quite faint, that is, if they are visible at all. These objects are faint for a variety of reasons: intrinsic faintness, diminutive size, obscuration by dust or debris or stars, or tremendous distance. And faint objects include not only galaxies but star clusters, emission nebulae, planetary nebulae, reflection nebulae, and supernovae remnants. And since the laws of physics work uniformly across orders of magnitudes of size as well as distance, the physics that describe a bright planetary nebula also describes a faint nebula.

In Chap. 6, we focused on a representative handful of facilities and modern astronomical surveys that are chasing down the faintest of objects, the likes of which had never been seen (nor have we had the technology to see them) until recently. We presented some information on NASA's "Four Great Observatories" in space, what they did, how they operated, and also looked at the up-coming James Webb Space Telescope mission.

In Chaps. 7, 8, and 9 we saw how one goes about observing faint objects successfully. It takes practice to be able to see some of these objects, but they are observable. Even the beginner with a desire to really "go deep" can benefit from the contents of these chapters. We present lists containing hundreds of objects in Chap. 10, and Appendices C and D give references to even more lists of objects. These lists cover objects of various levels of difficulty. But most can be successfully observed with a telescope in the 10- to 14-in. range under a reasonably dark, clear sky. There are also a wide variety of object types, from the mundane open cluster to the exotic quasar.

Chapters 11 and 12 show how one can stick to a faint objects observing program and even make contributions to the science of astronomy with one's faint object observations. These chapters provide many hints and a conclusion that ties its materials together. One more interesting project you can do after having observed faint objects for a while is to keep an inventory listing of all you have seen, organized by object name (or it in whatever order is meaningful to you, which can be by distance from Earth, date of observation, telescope used to observe it, etc.). That way you can see just how many unique objects you have personally observed and compare that number to the number your friend has observed. Then you can each compare your numbers to the actual number of objects that exist in the known universe and see just how insignificant is the amount of universe that you have actually observed! But every little bit of universe that you can "claim as your own" by personally observing and recording its image is well worth the effort.

This book has been made possible by combining the experience and expertise of a number of groups and individuals (see the "Acknowledgements" section) and is potentially greater than the sum of its parts. It is recommended that you become part of a community of like-minded people, if you have not done so already, and share your ideas and experience with one another, helping each other to grow and develop in the realm of amateur astronomy. This community can take the form of an astronomy club or special interest component of the astronomy club, or an on-line list serve, or some other mode of communication where members can share advice and information on new techniques, new objects, new experiences, etc., that the others could try. I hope that this book will provide enough material for these kinds of interactions to get started or to get carried to the next level.

Allow us to end on an observing note: This is a description, a narrative of a typical observing night at my own "dark site." Regularly (once per month, weather permitting) I make the 90-min drive from just northwest of Houston, Texas, to a site some 7 miles west of Columbus, Texas, to access some reasonably dark (Bortle Class 4) skies. I usually want to allow an hour or two after arrival before sunset so I can finalize my observing plan for the night, enjoy the lengthening evening shadows, the sunset, and any bright objects that first emerge from the fading twilight glow such as Venus, Jupiter, a colorful double star such as Albireo and the like. These put me in the mood for the long, busy, but enjoyable night ahead.

I have been, for the last several years, working on a number of different observing projects, but one project has been worked on steadily over the years: variable star magnitude estimates. I do most of this observing in my backyard, but I get the fainter stars at the site, where the 'scope is larger than what I use at home and the skies are much darker. I focus primarily on the stars that are designated by the AAVSO as "in need of more data," since I have a couple decades' worth of experience. These usually occupy roughly half of my observing time during the dark hours.

I also revisit old friends: the more prominent examples of deep sky objects in the skies. These are enjoyable to scrutinize on a regular basis. I may try something different, like boost the magnification up to 550× and look at bright planetary nebula at that power. I may compare the views an H-beta, a UHC, and an OIII filter give on the same emission nebula. Finally I go for the faint objects that are included in the program for the night. That program may include something like the Herschel 400 project, another Astronomical League observing project, or objects of my own choosing. A handful of faint objects that I look for each night are Solar System in nature, namely comets, which I am partially successful at actually seeing. The main cache of faint objects that I observe on site includes emission nebulae, planetary nebulae, galaxies, and supernova remnants. I like to look at a variety of objects and even some new objects (or obscure objects that I have not seen in a while). It is always enjoyable to scrutinize a faint object through the eyepiece while contemplating the physical nature of that object and the fact that this represents an infinitesimal fraction of light from this object that has traveled thousands, millions, even billions of light years to be intercepted by our planet and my telescope.

After working this way through the night, cycling through variable stars, then bright objects, then faint object, back to variable stars and repeat, the first gray signs of astronomical twilight appear too soon. This gives the "two-minute warning" (an American football expression that means time is almost up) to finish the last of my faint objects, before moving to brighter and brighter objects with the brightening sky. Finally the sky is bright enough so that only bright doubles and planets can be easily observed, and before one knows it, the daylight arrives in full brightness with the rising Sun bathing the treetops in its warm orange glow. Another enjoyable night has come to an end.

Appendix A

References and Footnotes

1. Wikipedia, the free encyclopedia (2012), History of astronomy, http://en.wikipedia.org/wiki/History_of_astronomy

2. M. Frommert, C. Kronberg(2005), Messier 45, http://seds.org/messier/Mdes/dm045.html

3. Praesepe Cluster, http://www.praesepe.com/praesepe_name.html

4. M. Frommert, C. Kronberg(2007), Messier 31, http://seds.org/messier/m/m031.html

5. M. Frommert, C. Kronberg (2006), Messier 42, http://seds.org/messier/Mdes/dm042.html

6. M. Frommert, C. Kronberg (2007), Messier 7, http://seds.org/messier/m/m007.html

7. M. Frommert, C. Kronberg (2007), Messier 39, http://seds.org/messier/m/m039.html

8. M. Frommert, C. Kronberg (1998), NGC 869 and NGC 884, http://www.seds.org/messier/xtra/ngc/n0869.html

9. Wikipedia, the free encyclopedia (2012), Coma star cluster, http://en.wikipedia.org/wiki/Coma_Star_Cluster

10. K.G. Jones (ed.), *Webb Society Deep-Sky Observer's Handbook*, vol. 3 (Open and Globular Clusters, Webb Society, 1980), pp. 3–4

11. K.G. Jones (ed.), *Webb Society Deep-Sky Observer's Handbook*, vol. 2 (Planetary and Gaseous Nebulae, Webb Society, 1979), pp. 8–9

12. Wikipedia, the free encyclopedia (2012), Nebula, http://en.wikipedia.org/wiki/Nebulae

13. Wikipedia, the free encyclopedia (2012), Globular cluster, http://en.wikipedia.org/wiki/Globular_cluster

14. Wikipedia, the free encyclopedia (2012), List of astronomical catalogues, http://en.wikipedia.org/wiki/List_of_astronomical_catalogues

15. K.G. Jones(ed.), *Webb Society Deep-Sky Observer's Handbook*, vol. 6 (Anonymous Galaxies, Webb Society, 1987), pp. 6–9

16. Wikipedia, the free encyclopedia (2012), Abell catalogue, http://en.wikipedia.org/wiki/Abell_catalogue

17. A. Ayiomamitis (2012), Abell planetary nebulae, http://www.perseus.gr/Astro-DSO-Nebulae-PN-Abell.htm

18. Courtesy of NASA, PGC Principle galaxy catalog (2003), http://heasarc.nasa.gov/W3Browse/all/pgc2003.html

19. Wikipedia, the free encyclopedia (2011), Principle galaxies catalogue, http://en.wikipedia.org/wiki/Principal_Galaxies_Catalogue

20. UGC—Uppsala General Catalog of Galaxies, Courtesy of NASA-GSFC (2004), http://heasarc.gsfc.nasa.gov/w3browse/galaxy-catalog/ugc.html

B. Cudnik, *Faint Objects and How to Observe Them*, Astronomers' Observing Guides,
DOI 10.1007/978-1-4419-6757-2, © Springer Science+Business Media New York 2013

21. L. Perek, L. Kohoutek (1967), Catalog of Galactic Planetary Nebulae. Academia publishing house of the Czechoslovak Academy of Sciences, Prague, p. 276

22. A. Hewitt, C. Burbidge (1993), A revised and updated catalog of quasi-stellar objects, http://w0.sao.ru/cats/doc/QSO2.html

23. Wikipedia, the free encyclopedia (2012), Sharpless catalog, http://en.wikipedia.org/wiki/Sharpless_catalog

24. Image source: http://upload.wikimedia.org/wikipedia/commons/a/a4/Charles_Messier.jpg

25. Sources (2012) http://en.wikipedia.org/wiki/Charles_Messier; Sea and Sky: Charles Messier, http://www.seasky.org/spacexp/sky5e06.html

26. F. Hartmut, C. Kronberg, G. McArthur, M. Elowitz, The SEDS Messier catalog webpages (http://messier.seds.org/) (SEDS, University of Arizona Chapter, Tucson, AZ, 1994–2000)

27. Reference: http://www.seasky.org/spacexp/sky5e06.html

28. Image source: http://en.wikipedia.org/wiki/File:William_Herschel01.jpg; {{PD-1923}} – published before 1923 and public domain in the US

29. M. Frommert, C. Kronberg (2008), William Herschel (1738–1822),http://seds.org/messier/xtra/Bios/wherschel.html

30. Image and information courtesy of Gary Kronk, http://cometography.com/biographies/herschelc.html

31. NNDB (2012), William Herschel, http://www.nndb.com/people/661/000096373/, Soylent Communications

32. M. Frommert, C. Kronberg (2012), Caroline Herschel (1750–1848), http://seds.org/messier/xtra/Bios/cherschel.html

33. Image Source: http://www.scientific-web.com/en/Astronomy/Biographies/images/JohnHerschel01.jpg

34. J.J. O'Conner, E.F. Robertson, *John Frederick William Herschel*, Credit: http://www-history.mcs.st-andrews.ac.uk/Biographies/Herschel.html (School of Mathematics and Statistics, University of St. Andrews, Scotland, 1999)

35. (2008) http://philosophyofscienceportal.blogspot.com/2008/04/sir-john-herschelastronomy-photography.html

36. J.A. Bennett (1990), John Lewis Emil Dreyer, Courtesy of Armagh Observatory, http://www.arm.ac.uk/history/dreyer.html, N. Ireland

37. Wikipedia, the free encyclopedia (2012), Edward Emerson Barnard, http://en.wikipedia.org/wiki/Edward_Emerson_Barnard

38. (2005) William Lassell Courtesy Mike Oats, http://www.mikeoates.org/lassell/lassell_inbrief.htm

39. Wikipedia, the free encyclopedia (2012), William Lassell, http://en.wikipedia.org/wiki/William_Lassell

40. Wikipedia, the free encyclopedia (2012), Edwin Hubble, http://en.wikipedia.org/wiki/Edwin_Hubble

41. Wikipedia, the free encyclopedia (2002), Edwin Powell Hubble (Biography), http://www.edwinhubble.com/

42. Wikipedia, the free encyclopedia (2012), Open cluster, http://en.wikipedia.org/wiki/Open_cluster

43. Image credit: NASA and the Space Telescope Science Institute and for European Space Agency (ESA) by the Hubble ESA Information Centre under Contract NAS5-26555.

44. This image file is licensed under the Creative Commons Attribution-Share Alike 3.0 Unported license Created by User: Worldtraveller {{GFDL}}

45. Wikipedia, the free encyclopedia (2012), Star cluster, http://en.wikipedia.org/wiki/Star_cluster

46. Wikipedia, the free encyclopedia (2012), Globular cluster, http://en.wikipedia.org/wiki/Globular_cluster

47. D.L. Chandler, *Sky and Telescope* (Sky Publishing Corporation, Cambridge, MA, 2002), September issue, p. 17

48. European Southern Observatory (19 Oct 2011), Astronomy News, *Astronomy Magazine*

49. Article and image credit: ESO/D. Minniti/VVV Team, Garching, Germany

50. Wikipedia, the free encyclopedia (2012), Emission nebula, http://en.wikipedia.org/wiki/Emission_nebula

51. M. Seeds, *Stars and Galaxies*, 6th edn. (© 2008 Brooks/Cole, a part of Cengage Learning, Inc., 2008). Reproduced by permission, www.cengage.com/permissions, pp. 203, 206 (Hereafter "SEEDS, pp…")

52. SEEDS, p. 206

53. Wikipedia, the free encyclopedia (2012), Planetary nebula, http://en.wikipedia.org/wiki/Planetary_nebula

54. SEEDS, p. 267

55. F. Kraljic, *Sky and Telescope* (Sky Publishing Corporation, Cambridge, MA, 2003), October issue, p. 112

56. SEEDS, p. 212

57. Wikipedia, the free encyclopedia (2012), Supernova remnant, http://en.wikipedia.org/wiki/Supernova_remnant

58. SEEDS, p. 281, 282

59. Wikipedia, the free encyclopedia (2012), List of supernova remnants, http://en.wikipedia.org/wiki/List_of_supernova_remnants, available under the Creative Commons Attribution-ShareAlike License

60. {{PD-US}} – published in the US before 1923 and public domain in the US

61. M. Seeds, *Stars and Galaxies*, 6th edn. (© 2008 Brooks/Cole, a part of Cengage Learning, Inc., 2008). Reproduced by permission, www.cengage.com/permissions, pp. 343–344 (Hereafter "SEEDS, pp…")

62. Wikipedia, the free encyclopedia (2012) Galaxy, http://en.wikipedia.org/wiki/Galaxy

63. The Hubble Tuning Fork on the website of the Sloan Digial Sky Survey/Sky Server, http://cas.sdss.org/dr3/en/proj/advanced/galaxies/tuningfork.asp

64. SEEDS, pp. 345–346

65. The Hipparcos space astrometry mission, http://www.rssd.esa.int/index.php?project = HIPPARCOS

66. J. Lucentini, The mystery of galaxy spirals. *Sky and Telescope*, (Sky Publishing Corporation, Cambridge, MA, 2002), September issue, pp. 36–44

67. SEEDS, p. 353

68. Wikipedia, the free encyclopedia (2012), Elliptical galaxy, http://en.wikipedia.org/wiki/Elliptical_galaxy

69. Wikipedia, the free encyclopedia (2012), Local group, http://en.wikipedia.org/wiki/Local_Group

70. SEEDS, pp. 358–360

71. SEEDS, p. 356

72. Wikipedia, the free encyclopedia (2012) Galaxy groups and clusters, http://en.wikipedia.org/wiki/Clusters_of_galaxies

73. Wikipedia, the free encyclopedia (2012) Bullet cluster, http://en.wikipedia.org/wiki/Bullet_Cluster

74. Wikipedia, the free encyclopedia (2012) Active galactic nucleus, http://en.wikipedia.org/wiki/Active_galactic_nucleus

75. From SEEDS, *Stars and Galaxies*, 6th edn. (© 2008 Brooks/Cole, a part of Cengage Learning, Inc., 2008). Reproduced by permission, www.cengage.com/permissions, pp. 368–369

76. Wikipedia, the free encyclopedia (2012) Quasar, http://en.wikipedia.org/wiki/Quasar

77. Wikipedia, the free encyclopedia (2012) Blazar, http://en.wikipedia.org/wiki/Blazar

78. Wikipedia, the free encyclopedia (2012), BL lac object, http://en.wikipedia.org/wiki/BL_Lac_object

79. Image source: http://hubblesite.org/newscenter/archive/releases/1996/35/image/a/format/small_web/

80. Wikipedia, the free encyclopedia (2012) Keck telescope, http://en.wikipedia.org/wiki/Keck_Telescope
81. Wikipedia, the free encyclopedia (2012), Hubble space telescope, http://en.wikipedia.org/wiki/Hubble_Space_Telescope
82. The Telescope: Hubble essentials (2012) http://hubblesite.org/the_telescope/hubble_essentials/
83. Wikipedia, the free encyclopedia (2012), Chandra x-ray observatory, http://en.wikipedia.org/wiki/Chandra_X-ray_Observatory
84. Wikipedia, the free encyclopedia (2012), Spitzer space telescope, http://en.wikipedia.org/wiki/Spitzer_Space_Telescope
85. Wikipedia, the free encyclopedia (2012), Compton gamma ray observatory, http://en.wikipedia.org/wiki/Compton_Gamma_Ray_Observatory
86. Wikipedia, the free encyclopedia (2012) Gran Telescopio Astronomical, http://en.wikipedia.org/wiki/Gran_Telescopio_Canarias
87. Wikipedia, the free encyclopedia (2011) Astronomical survey, http://en.wikipedia.org/wiki/Astronomical_survey
88. (2012) About the Webb, http://www.jwst.nasa.gov/about.html
89. Wikipedia, the free encyclopedia (2012), James webb space telescope, http://en.wikipedia.org/wiki/James_Webb_Space_Telescope
90. W. Steinicke, R. Jakiel, *Galaxies and How to Observe Them*, 1st edn. (Springer, New York, 2007), pp. 92–96 (Hereafter "*Galaxies…*, pp…")
91. Some of the mirror, lens, and eyepiece cleaning tips obtained from various individuals, personal experience and the website: http://www.sctscopes.net/SCT_Tips/Maintenance/Cleaning_Your_Optics/cleaning_your_optics.html
92. Wikipedia, the free encyclopedia (2012) Table source, http://en.wikipedia.org/wiki/Bortle_Dark-Sky_Scale
93. *Galaxies…* adapted from Table 5.1, p. 93
94. Lumicon Website = http://www.lumicon.com/astronomy-accessories.php?cid = 1&cn = Filters
95. *Galaxies…*, pp. 84–86
96. *Galaxies…*, pp. 77, 88
97. *Galaxies…*, pp. 89–90
98. Ideas inspired by *Galaxies and How to Observe Them* Steinicke, Wolfgang; Jakiel, Richard; 1st edn, Springer 2007, pp. 103–108 (Hereafter "*Galaxies…*, pp…")
99. *Galaxies…*, pp. 105–106
100. *Galaxies…*, pp. 106–108
101. G. North, *Advanced Amateur Astronomy* (Edinburgh University Press, Edinburgh and New York, 1991), pp. 265–267
102. D. Pendick (2009), Profile: Youngest person to discover a supernova, *Astronomy* Magazine. Online News Article, at http://www.astronomy.com/News-Observing/News/2009/06/Profile%20Youngest%20person%20to%20discover%20a%20supernova.aspx
103. Wikipedia, the free encyclopedia (2012), Citizen science, http://en.wikipedia.org/wiki/Citizen_science

Appendix B

Glossary

Abell catalogue   A catalog of 2,712 rich galaxy groups and planetary nebulae constructed by the astronomer George O. Abell; he also cataloged more than 80 planetary nebulae found on Palomar Sky Survey plates.

Absolute magnitude   The apparent magnitude (brightness) of a star if it were placed at 10 parsecs (32.6 light years) distance.

Absorption line spectrum   Spectrum consisting of dark lines against a bright continuum background; occurs when light from a continuum source passes through a cool thin gas.

Accretion disk   A disk of material that encircles a super massive black hole in an active galaxy. The material in the accretion disk feeds the black hole, causing the copious amount of radiation that makes the object visible from great distances. Accretion disks are also found in close binary systems where one component overfills its Roche lobe and spills material toward the companion, which either receives it directly or collects it into a disk of material (the accretion disk).

AGN   Active Galactic Nuclei, strong radio sources, thought to arise from the act of a super massive black hole consuming material at the center of a normal galaxy.

Antonin seeing scale   The rating system that gauges the seeing (steadiness) of the atmosphere at a given observing site, using the convention that I is the steadiest and V is the worst, and II, III, IV are intermediate values.

Apparent magnitude   The brightness of an object gauged on a logarithmic scale that was originally devised by the ancient Greek astronomer Hipparchus and is still used (with modification) today.

Astronomical seeing   A measure of the steadiness of an astronomical object; excellent seeing means that stars twinkle very little and fine planetary detail is seen.

Averted vision   The practice in faint object astronomy where the observer does not look directly at an object but slightly off center to enable the light from the object to fall upon the rod cells of the eye, making the object more visible to the observer.

B. Cudnik, *Faint Objects and How to Observe Them*, Astronomers' Observing Guides, DOI 10.1007/978-1-4419-6757-2, © Springer Science+Business Media New York 2013

Barnard, Edward E. American astronomer who developed photographic techniques to survey the heavens and discovered (along with Max Wolf) the true nature of dark nebulae; he also made several key astronomical discoveries.

Blazar A very compact quasar at the center of an active, giant elliptical galaxy. These objects are among the most energetic known and are a type of Active Galactic Nucleus.

BL Lacertae objects A type of active galaxy with an active galactic nucleus, which is characterized by rapid, large scale variability in flux. It also exhibits significant (up to a few percent) optical polarization.

Bortle light pollution scale A standard reference for judging the quality of the nighttime sky in terms of darkness, where a Bortle 0.5 sky is about as dark as they get on Planet Earth, and a Bortle 9 signifies severe light pollution.

Catalog (star) A listing of objects organized to include relevant details such as celestial coordinates, apparent magnitude, apparent size, and other information.

Cepheid variable star A special kind of variable star that shows a pulsation period to luminosity relationship that enables astronomers to estimate distances to other galaxies. Edwin Hubble used these stars to show that the Andromeda Galaxy was a vast star system like the Milky Way, rather than a spiral nebula within the Milky Way.

Chandra X-ray observatory A space telescope designed to observe in many "colors" within the X-ray part of the electromagnetic spectrum; it was launched in July 1999 and has returned images that have led to significant advances in X-ray astronomy.

Citizen science Umbrella term signifying a family of activities that ordinary citizens can do to help in scientific research, from bird watching to sunspot counting; there are numerous examples in this expanding area of science.

Compton gamma ray observatory One of NASA's four "Great Observatories," which observes in the hard (shortest wavelength/highest energy) X-ray and gamma ray portion of the electromagnetic spectrum. During its 9-year mission (1991–2000) it has advanced our knowledge of enigmatic gamma ray bursts and has also illuminated our views of objects closer to home, such as supernova remnants and energetic solar flares.

Dreyer, John L. E. Danish astronomer who lived during the latter part of the nineteenth and early twentieth century and was a key figure in the development of the *New General Catalogue of Nebulae and Clusters of Stars*.

Electron volt (eV) Unit of energy used especially for atomic and nuclear physics; it is defined as the energy given to an electron by accelerating it through 1 V of electric potential difference; it is equivalent to 1.6×10^{-19} J. A keV is 1,000 eV and an MeV is 1,000,000 eV.

Elliptical galaxy Generally dust-poor galaxies consisting of mostly ancient stars and no spiral structure; these range in shape from round to elliptical.

Emission line spectrum   Spectrum consisting of bright lines against a faint or dark continuum and occurs when ionized atoms or molecules of a cloud of gas return to lower energy levels, releasing photons of specific energy and color.

Emission nebula   Otherwise known as an HII region, these nebulae are luminous patches of gas that shine through ionization and recombination. The source of the energy is the ultraviolet radiation from nearby hot O- and B-type stars. The Orion Nebula (M42) is the brightest and nearest example of an emission nebula.

Exit pupil   The amount of light coming from the telescope, whose size depends on the instrument's parameters only (aperture and focal length).

Faint objects   General descriptor that includes any astronomical object near the low end of visibility and can include any Solar System object or deep sky object that takes at least some effort to see.

Galaxy   A vast system of stars, gas, and dust that make up a system; there are an estimated 100 to 200 billion such systems, great and small, in the visible universe. These are classified by their appearance in three main types: spiral (barred and regular), elliptical, and irregular.

Giant Magellan telescope   Currently in the preliminary stages, this telescope is expected to take the title of "largest ground-based optical telescope" when it becomes operational sometime around 2020. It will be based in South America, in Chile's Atacama Desert, and will have seven 8.4-m mirrors working together to form a single large mirror 24.5 m (80 ft) in diameter.

Globular star clusters   Spheroidal groupings of stars, from several thousand to over one million, which inhabit the halo as well as the disk of the Milky Way and other galaxies.

Gran telescopio canarias   This is currently (early 2012) the largest ground-based single mirror unit optical telescope in the world, operating on one of the Canary Islands. Its primary is a composite of hexagonal units that have a total diameter of 10.4 m (410 in.). The telescope saw first light in 2007 and has operated at full aperture since 2009.

Herschel, Caroline   Sister of William Herschel; assisted him with his astronomical observations and made some discoveries of her own.

Herschel, John   Son of William Herschel, John carried on his father's work and included the Southern Hemisphere sky as he continued the survey from South Africa, adding 1,700 objects.

Herschel, William   Eighteenth-century astronomer who surveyed the night sky and cataloged a total of 2,500 objects. He also discovered the planet Uranus and made other important contributions to astronomy.

Hipparchus   Ancient Greek astronomer who cataloged the stars and is credited for inventing the stellar magnitude system.

HIPPARCOS Satellite (HI Precision PARallax COllecting Satellite) whose primary mission was to measure the parallax of over one million stars with some degree of precision during its 3.5-year mission in the early 1990s.

Hubble, Edwin P. American astronomer who discovered the expansion of the universe, formulated a classification scheme for galaxies in the form of a tuning fork, and was instrumental in defining the extragalactic nature of "spiral nebulae."

Hubble space telescope (HST) The most popular orbiting observatory, HST was launched by the United States in 1990 and continues to operate to this day. It has returned hundreds of thousands of stunningly beautiful and scientifically important images and has greatly advanced our knowledge of the cosmos.

Hyades V-shaped open star cluster, one of the brightest in the nighttime sky and easily visible with the naked eye.

Index catalogue (IC) Supplement to the *New General Catalogue* and is presented in two parts: *Index Catalogue I* and *Index Catalogue II*.

James Webb space telescope (JWST) Formerly the NGST (Next Generation Space Telescope), this successor to the HST will be much larger (primary mirror diameter of 6.5 m compared to 1.8 m for HST's primary mirror) and will operate primarily in the infrared part of the spectrum. The launch is expected to occur in 2018, and the telescope is expected to return unprecedented images almost to the very edge of the visible universe, near the time of recombination.

Keck observatory A pair of the world's largest telescopes is housed on the summit of Mauna Kea, each with primary mirrors of 10 m (33 ft) in diameter. This observatory features adaptive optics, which provides images with a quality that at least matches that from the Hubble Space Telescope.

Lassell, William Nineteenth-century English astronomer who made several astronomical discoveries, including the discoveries of Triton at Neptune in 1846, Hyperion at Saturn in 1848, and Ariel and Umbriel at Uranus in 1851.

Lenticular galaxy A type of galaxy that appears as a transition between elliptical galaxies and spiral galaxies. These objects are so-named because of their superficial resemblance to convex lenses. They contain a central bulge, gas and dust, but no spiral arms. Some of them display a central bar, like a barred spiral galaxy.

LPR filter Light Pollution Rejection filter, designed to reduce the glow of light pollution from artificial sources, but designed to allow wide bandpasses of light from natural emission sources (OIII, H-beta, H-alpha).

Messier, Charles Eighteenth century astronomer who cataloged 107 deep sky objects that now bear his name (three have been added posthumously to make a grand total of 110).

Narrowband imaging Imaging deep sky objects with filters of SII (Ionized Sulfur), OIII, H-alpha in place of the traditional tri-color setup; this makes use of these primary components of nebulae as well as mapping out where in the object these dominate.

Nebulae   Clouds of interstellar material classified by their physical nature: emission (glow by the ionization/recombination of component atoms stimulated by ultraviolet radiation from nearby hot stars), reflection, planetary (appearing disk-like or ring-like, the end product of the evolution of sun-like stars), supernova remnant, and dark (appearing in silhouette against bright backgrounds of either emission nebulae or star clouds).

New general catalogue (NGC)   This is one of the most widely used by astronomers worldwide, contains 7,840 entries.

OIII filter   A filter, which allows the light from the emission of twice-ionized Oxygen (OIII) to pass while blocking the rest; this is often used to enhance the visibility of certain planetary and emission type nebulae.

Open star clusters   Grouping of stars, ranging from a half dozen up to several thousand, that are gravitationally bound, and formed together near the disk of the Milky Way.

Optically violent variable (OVV)   A type of blazar or powerful quasar that varies dramatically over very short timescales.

Perek-Kohoutek catalog (PK)   A catalog of 1,510 galactic planetary nebulae.

Pleiades   Also known as Messier 45, one of the brightest open star clusters visible in the nighttime sky; is easily seen with the naked eye.

Principal galaxies catalogue (PGC)   A catalog of 73,197 galaxies, combining the contents of several other catalogs of galaxies. A second version of this was published in 1996 and contains 108,792 entries.

Quasars   Also known as "Quasi-Stellar Objects," faint star-like points of light with peculiar emission spectra.

Recombination (the time of...)   The time in the early history of the universe when the temperature cooled sufficiently to allow free electrons to combine with free protons and helium nuclei to form hydrogen and helium atoms. This was the moment that the universe transitioned from being opaque to being transparent and marks the theoretical limit to our view of the distant universe.

Reflection nebula   A nebula that shines by the reflected light of a nearby star or stars. M78 in Orion is a bright example, as is part of the Trifid Nebula (M20) in Sagittarius.

RR Lyrae variable star   A special class of pulsating variable star that shows a relationship between the pulsation period and luminosity. These are fainter than Cepheid variables and tend to be spectral class A (compared to F or G for Cepheid variables); these were useful in finding distances to globular clusters within the Milky Way and within the nearest galaxies.

Sky transparency   A measure of the clarity of the nighttime sky. A transparent sky is one that allows the maximum amount of visible light from extraterrestrial sources; this light is little dimmed by haze, light pollution, aerosol pollution, and clouds.

Spectral type   The classification of stars by their spectrum, which is related to their surface temperatures. The spectral class scheme runs like this: O,B,A,F,G,K, and M, with the O-types the hottest (with the surface temperature in excess of 20,000 K) and the M-types being the "coolest" (at just above 2,000 K surface temperature). The Sun is right of center, with a G2 spectral type and a surface temperature of 5,800 K.

Spectroscopic parallax (with distance modulus and main sequence fitting)   An astronomical method for measuring distances to the stars that can be used for any main sequence star whose spectral type and apparent magnitude can be determined. Once the spectral type is determined, the absolute magnitude can be estimated by theory and the distance modulus formula (m−M = 5 log(d/10pc), used to determine distance to the star). This is used for single stars. A related method, called "Main Sequence Fitting," is used for star clusters; in these, the cluster members are plotted on an H-R diagram using each member's spectral type and apparent magnitude. The vertical displacement of the main sequence of the cluster gives m−M, which then can be used to calculate the cluster's distance (with the valid assumption that all cluster members are more-or-less at the same distance).

Spiral galaxies   Vast systems of stars, gas, and dust assembled in a disk-shaped formation with spiral arms, a central bulge, and, in some cases, central barred structures (for barred galaxies).

Spitzer space telescope   The fourth and final of NASA's "Great Observatories," this space telescope was launched in 2003 and observes the cosmos in the infrared part of the spectrum. It continues to operate to this day, but only in the shortest wavelength part of the IR, due to the depletion of liquid helium (essential for the operation of instrumentation to observe in longer IR wavelengths) on May 15, 2009.

Stellar association   A group of stars that form together but are not gravitationally bound; these appear similar to open clusters but are generally younger.

Supernovae   Among the brightest explosions in the nighttime sky, these occur when either a high-mass star reaches the end of its life and collapses upon itself, triggering an explosion (Type II), or when a white dwarf star in a binary system accretes enough mass to push it over the Chandrasekhar limit of 1.46 solar masses, causing the star to detonate (Type I).

Surface brightness   A measure of how bright (or faint) an object appears; is expressed in units of magnitudes per square arc minute or per square arc second.

Tidal radius (of a globular star cluster)   This is the distance in a cluster of stars where the external gravity from the host galaxy has more influence on the cluster stars than the cluster itself.

Trigonometric parallax   Sometimes called "parallax" or "parallax method"; this is the technique used to measure the distances to the nearest stars through triangulation, which happens when the star of interest is measured with Earth at one end of its orbit, then measured again 6 months later. At this point the star has exhibited a shift in position due to its relative nearness and our changing perspective.

UHC filter   Ultra High Contrast filter that is designed to enhance the contrast of emission and planetary nebulae.

Ultra deep field (UDF) image   Famous image of the most distant parts of the universe, featuring tens of thousands of never-before seen galaxies captured by the Hubble Space Telescope. This more than 30-hour exposure captures some of the most distant objects ever imaged.

Variable stars   Stars that change in brightness over time, due to their actual change in brightness (usually by pulsation) or size or their being blocked (partially or totally) by a companion.

Appendix C

Resources Useful for the Observation of Faint Objects

What follows are several lists that should be useful for your observing experience. We start with a checklist that you should keep (modify it for your own needs) and follow each time you head out to a dark site location. This will ensure that you will not forget important small items. Next are a few websites that are useful in determining the quality of the sky during a planned observing run. There are also a few suggestions if you want to get into astrophotography and narrow band imaging, and a few sites that will provide information on cleaning the eyepieces properly, where to get your mirrors aluminized, and help on collimating your telescope. The information in this section is meant to get you started and is by no means a complete listing. As has been mentioned a few times, there are resources in the local astronomy club, online, and elsewhere that can help you with whatever your astronomical need may be.

Observing Checklist

Have this checklist handy and run through it as you are getting your things together for an observing run, whether it is one night or several. You may not need to include everything listed below, or you may have additional things not included here, but be sure you have such a checklist and use it all the time.

I. Telescope
1. Primary mirror if removable
2. All the parts that go with the 'scope, including base, struts, top, shroud, and all the nuts, bolts and screws (and some extras in the event some are lost…) to hold the 'scope together
3. For a Celestron or Meade, both the tube and mount, and any attachments that go with these
4. Finderscope(s)
5. Ladder, if it is a tall "Obsession-type" Dobsonian

B. Cudnik, *Faint Objects and How to Observe Them*, Astronomers' Observing Guides, DOI 10.1007/978-1-4419-6757-2, © Springer Science+Business Media New York 2013

6. Collimation kit, laser collimator

7. Gloves to handle the mirror if needed

II. Eyepieces
  1. Three or more eyepieces of different focal lengths to include wide field, low power, higher magnification, etc.
  2. Canned air to blow dust off of mirrors and eyepieces
  3. III, UHC, LPR, or other filter needed to enhance the visibility of your targets
  4. Hair dryer to dry dew that collects on telescope optics

III. Tarps
  1. One to cover the ground beneath the telescope (unless your site has specially-built concrete pads)
  2. Additional tarps and other plastic to cover and shelter the telescope by day if going for multiple days
  3. Many bungee straps or other devices to hold tarps/coverings in place
  4. Stakes and a hammer to pound them into the ground

IV. Accessories and Miscellaneous
  1. Table (to keep computers and other accessories)
  2. Power cord and at least 100 ft of additional extension cord if the site has electricity and your 'scope requires it
  3. Observing chair and other chair to rest in
  4. Red and white light flashlights and extra batteries for these lights
  5. Binoculars
  6. Laptop with red screen; make sure its battery is fully charged before you leave unless you will plug it into an AC power source on site
  7. Snacks, water, non-alcoholic drinks
  8. A plan of observation for the night(s) you will be observing
  9. Insect repellant (spray yourself WELL AWAY from scopes and don't get the stuff on your fingers)
  10. Tent, sleeping bag, camping accessories if you will be camping
  11. An alternate plan or other things to do in case you get clouded out

Determining Sky and Weather Conditions

The following are websites that are helpful in determining the expected quality of the sky on a night when planning to go observing. Note that these are not the only sites to find this information but are a good representation of what's available.

1. www.weather.gov is the National Weather Service website. It has a field where you can enter your city and state and it will take you to the website of the local forecast office. You can view the forecast and look for the link to the forecast discussion page to get more information than what is given in the standard forecast.

2. http://cleardarksky.com/csk/ is the Clear Sky Chart homepage, a very useful tool for getting the astronomically useful weather forecast. You can use this if you live in the United States, Canada, Mexico, or the Bahamas. This provides forecasts for up to 48 h for cloud cover, seeing, transparency, humidity, and more.

3. http://www.weatheroffice.gc.ca/astro/clds_vis_e.html is another site that provides forecasts for cloud cover over a large area (eastern and western United States, eastern and western Canada).

4. http://7timer.y234.cn/ "7Timer" numerical weather forecasts for anywhere in the world

5. http://philhart.com/content/cloud-forecasts-astronomers is another weather resource geared toward astronomical purposes.

6. http://www.skippysky.com.au/ Skippy sky provides a wealth of information out to several days to include low cloud coverage, high cloud coverage, seeing, transparency, humidity, and more.

7. http://www.wunderground.com/ provides forecasts for locations worldwide.

In addition one can go to the website http://aa.usno.navy.mil/index.php, the Astronomical Applications Department of the U.S. Naval Observatory, to get sunrise/sunset and moonrise/moonset times and other astronomical information.

Narrowband and Broadband Astro-Imaging

Books about astro-imaging can be found in many places, and there are websites that have links to electronic books that provide valuable information for the beginner to get started with the activity, both narrowband and broadband. The following are web sites and book information to get you started.

1. http://www.astropix.com/HTML/SHOW_DIG/Elephant_Trunk_Nebula.HTM starts with a nice picture of the Elephant Trunk nebula and it links to two e-Books (on CD-ROM) about astrophotography with digital cameras.

2. http://www.narrowbandimaging.com/home_page.htm also has some useful resources and sample images to get one started with narrowband imaging

3. www.amazon.com has links to many books on astrophotography, you can go to: http://www.amazon.com/s/?ie=UTF8&keywords=astrophotography+ books&tag=googhydr-20&index=stripbooks&hvadid=15115817755&ref=pd_ sl_6iwuh47uzw_b

One of the best sources of narrowband imaging is Don Goldman (astrodon. com) and the answers to the most frequently asked questions concerning narrowband imaging can be read at http://www.astrodon.com/Orphan/astrodonfaqnarrowband/. He lists all of the scientific data regarding his filters and a few others, and explains the need to understand the camera's sensitivity and exposure rates.

Sketching What You See: Step-by-Step Guides

There are a number of books and resources on the market today that show the interested astronomer how to make quality drawings of what they see at the eyepiece. One can find how-to resources on the *Astronomy* and *Sky & Telescope* magazine web pages. Also, websites like http://www.backyard-astro.com/index.html offer resources and links to resources on how to get started (and keep going) in this wonderful hobby.

The website for *Astronomy* magazine is www.astronomy.com; for *Sky & Telescope* magazine it is www.skyandtelescope.com. Note that some features on these websites require the user to have a subscription with the magazine in order to use these features. *Astronomy* Magazine's sketching tutorial is one such feature. The magazines are certainly worth the money paid for the subscription and they (especially *Sky & Telescope*) have regular articles featuring lists of faint (and not so faint) objects.

Also the following books (available on websites such as amazon.com) have some useful material to show you how to make quality astro-sketches:

- *Astronomical Sketching: A Step-by-Step Introduction* (Patrick Moore's Practical Astronomy Series), by Richard Handy, David B. Moody, Jeremy Perez, Erika Rix, and Sol Robbins, 2007
- *How to Use an Astronomical Telescope* by James Muirden, 1988.

Accessories to Care for Your Equipment

For mirror re-aluminization http://www.amateurtelescopemaker.com/mc.htm, which is the Amateur Telescope Maker's Resource List (also a resource list for the average amateur astronomer). This serves not only telescope makers but also amateur opticians, and amateur astronomers in general.

Cleaning Your Optics and Eyepieces

- http://www.sctscopes.net/SCT_Tips/Maintenance/Cleaning_Your_Optics/Cleaning_Optics_with_Compressed_Air.txt
- You may benefit from a CO_2 duster kit, item #114 0208 sold on X-tremeGeek.com for $37 as of this writing. Or you can get a can of "Velocity" dust remover from Radio Shack. http://www.sctscopes.net/SCT_Tips/Maintenance/Cleaning_Your_Optics/cleaning_your_optics.html

Eyepiece cleaning: Should be done very carefully and as seldom as possible. Televue has some excellent guidelines for cleaning eyepieces and binocular optics. Go to their website: http://www.televue.com/engine/page.asp?ID=143.

Telescope collimating: This can be quite a challenge, speaking from experience. I will include a couple of resources here to help find information on how to do the collimation as well as some suggested accessories for collimation.

http://www.andysshotglass.com/ is a website that comes highly recommended from experienced amateur astronomers. To access the collimation tutorial, go to the website, click on "articles," click on the "Do it Yourself" link, then look for "collimation." http://www.astro-tom.com/telescopes/collimation.htm also contains an excellent tutorial on how to collimate your Schmidt-Cassegrain or Newtonian reflectors.

Some Recommended U.S. Vendors

Whether you are in the market for a new or larger telescope, need accessories (including eyepieces or cameras), or just like to shop, these websites link you to some vendors that provide quality products at reasonable prices.

- Celestron telescopes and more: http://www.celestron.com/
- Meade telescopes and more: http://www.meade.com/
- For binoculars, try: http://www.bushnell.com/
- Many types of equipment and accessories: http://www.opticsplanet.net/
- Filters for deep sky observing: http://www.lumicon.com/
- Oceanside Photo and Telescope (San Diego, California): http://www.optcorp.com/
- Land, Sea, and Sky (Houston, Texas): http://www.landseaskyco.com
- The Santa Barbara Instrument Group: http://www.sbig.com/
- Astronomics: http://www.astronomics.com/

Appendix D

Observing Software

What follows is a short listing and brief description of several examples of popular software being used by amateur astronomers at the time of this book's writing. At the end each description is a web address where the reader can get more information and possibly make a purchase. The last software is the featured software for faint objects.

AstroNotes

Go to http://astronotes.ca/ to download this free archiving software. There is more information, screenshots, and files to download. This is freeware that anyone can download and try out.

C2A

Here is another example of freeware that one can download and try out. It loads quickly and supports a myriad of star and deep sky catalogs. It is intuitive and easy to use. This software has recently been fully updated and the manual is written in English (the author is French). It prints out great star charts. Visit http://www.astrosurf.com/c2a/english/information.htm to learn more and to download the software.

Cartes du Ciel

Another example of astro freeware is *Cartes du Ciel*. It works cross platform (Win, Mac, Linux) and comes with multiple display modes for stars and deep sky objects and tons of other configuration options. It can use star catalogs such as the UCAC3 and older ones, the various GSC versions, Tycho2, etc. Easily available catalogs for deep-sky objects—NGC, RNGC, PGC, GCM, GPN, LBN, OCL, etc., and you can easily generate your own from other sources. It can pull images from the *Digital Sky Survey* or *RealSky* in multiple formats, has telescope controls, and there are several

B. Cudnik, *Faint Objects and How to Observe Them*, Astronomers' Observing Guides, DOI 10.1007/978-1-4419-6757-2, © Springer Science+Business Media New York 2013

other applications that can interface with it such as *Deepsky Astronomy* software. You can get it at http://www.ap-i.net/skychart/.

Deepsky Astronomy

http://www.deepsky2000.net/whatis.htm has lots of information about this suite of observational software tools, including demos and testimonials.

MaximDL

This is a suite of software to meet any observatory's needs. It provides a fast and easy way to take images of the nighttime sky and is useful for applications ranging from casual astro-imaging to collecting data of scientific quality. It is useful for imaging, image processing, and calibration, and its Version 5 adds many new capabilities. Go to http://cyanogen.com/ to read more about this and to find out about other tools they offer such as cloud sensors, MaxPoint™ to ensure your telescope always points accurately, Quick Fringe, and more.

Megastar

The program *Megastar* is a popular one for planning observing sessions. In some cases, this is the sole purpose that people use the program. One can download comet elements and enter them into *Megastar* so as to be able to determine where a comet is at any time, including the RA and Declination. This is useful for the GOTO function of such telescopes, and the altitude and azimuth given indicates whether it is visible or not.

This can be done with other objects and *Megastar* has almost all catalogs available. It also contains images of many objects as well. *Megastar* also offers many more capabilities that come in handy from time to time.

For more information, and for a free download to try out the program before buying it, go to the website http://www.willbell.com/software/megastar/download.htm.

Starry Night

This software comes in different experience levels, from backyard to professional. One can use the professional version to generate finder charts as needed. You can also use this program as instructional software. It is often provided with astronomy textbooks and has outstanding graphics and real time and/or time lapse animations. You can pretend you are in a spaceship and take a trip outside the Solar System or to a planet, where great renderings make it seem like you are really there.

The software shows the names of the craters on the Moon and accurately simulates light pollution from nearby cities if you opt to show it. It also provides the sounds of the nighttime to make you feel like you are really outdoors under the stars. The daylight sky looks realistic, and you can use a variety of "horizons," including your own custom horizon, for your scenery. It is a little less intuitive than some of the others on the market, but it is a great program overall.

For more information go to www.starrynight.com.

Stellarium

This is some excellent planetarium-style software that can determine whether a twilight variable star target is accessible during a given night. *Stellarium* is free software and is useful not only for the planetarium aspect but to drive telescopes. For example one user makes use of this software to drive his CG5 GOTO and his old Meade ETX125ec.

One can simulate the night time sky under a wide variety of conditions, with or without atmosphere. One can control the twinkling of the stars, and even remove the ground to see what constellations are underfoot! You can go anywhere on Earth with this program and see what the sky looks like at any given time. There are 63 worlds you can view the night sky from, including the Sun and Comet C/2006 P1 (McNaught). These worlds include about a half dozen major asteroids, all the major planets and a few of the dwarf planets, and many of the natural satellites that orbit the planets of the Solar System. There are also three other worlds you can select scenery from—the Moon, Mars, and Saturn. The rest of the scenery images are from different spots on Earth.

So if the skies are cloudy one night, *Stellarium* makes for some excellent cloudy night activity. You can browse what the night sky would have looked like from your location that night, or go to Mars, or Saturn, and see the dance of the many moons in these planets' skies. Go to www.stellarium.org to download a copy for yourself.

Sky 6/Sky X

This program features a realistic simulated view of the sky and a host of objects displayed. It is used to control the C14 and can point at one such object, click on the object, and tell the telescope to slew to the object and you are there. You can zoom in until you get an FOV comparable to that in the eyepiece, and click on a star for an identification and a magnitude. You can also click and drag from one star to another to get a measure of the separation between the two. *TheSky* 6 also features a searchable database of NGC, Messier, and IC objects, so that if one knows the number of the object, one can type this in and do a search for the object and, if it is above the horizon, slew to that object. For variable stars and more obscure objects (including comets), you are able to easily enter the coordinates and either center the location on the position entered or get the telescope to slew to that position.

One astronomer cautions, when upgrading from Version 6 to Version X, to beware. This transition was a big one for him, not just because of the price of the program but because it was much like learning a whole new program. One drawback is the removal of the page up and down zoom features, which was useful in the dark (and the "draw a box," click, and zoom doesn't work that way anymore). Another pitfall is the potential of a lot of CPU/battery usage if one is not careful with one's refresh settings. This same astronomer offers a suggestion that helps him cope with these huge changes: use the *Sky X* for planning and researching what will be and has been observed, and continue with the *Sky 6* for running everything out in the field and under the stars.

Another astronomer mentions that this is one of the more expensive software, but it does well with GOTO telescope mounts. It also works with dome control software and supports a myriad of star and deep sky catalogs. It has a useful add-on program called T-Point, which compensates for all the deflections and irregularities in mount driving and pointing, so one always gets objects in the field of view, even the FOV of one's camera. This is a great professional grade program but the least intuitive.

More information can be found at http://www.bisque.com/sc/pages/thesky6-family.aspx.

Featured Product: Sky Tools 3

Following is an overview of the standard edition and tools useful for observing faint objects. Lots more information can be found at http://www.skyhound.com/ but we will spend a little time here describing the usefulness of the software. This is an excellent program for planning observing sessions. You can customize the planning for your telescope, your fields of view, and your observing site with its limiting magnitude. The planning tool helps you to know when the best time to see each object is, and it orders the objects so that those that become well placed earlier are displayed earlier, those that are well placed later are displayed later, etc. If an object isn't at its best from your location on a given night, you can filter it out, so you can concentrate on those objects that will be at their best.

The planning tool also includes basic object information such as constellation, apparent magnitude, apparent size, and the physical distance to the object, if known. It also has information on the sizes of the Barnard dark nebulae. You can customize the display with any of a large number of columns provided for visual observing. For comparison, the visual difficulty under best conditions is provided, and the best eyepiece to use for each object is also given. You can also use other columns to prioritize objects, check whether or not the object has been observed and a five-star type rating of the object. You can also go to log entries for the object and make new entries in the field as you observe.

If you cannot think of what you want to observe, no problem! The skyhound website has over 200 ready-made observing lists that people can download and use. If you have an observing list that you would like to share with others, you can do that also. There is a community of *Sky Tools* users that share observing hints, lists, and experiences and is useful for anyone who is interested in going to the next level in observational astronomy.

Sky Tools comes in three editions: the starter edition, the standard edition and the pro edition. The novice starting out in astronomy for the first time can benefit from the starter edition, which helps beginners get started with and learn about observational astronomy. There is a free trial version of this edition that one can download and try out and see whether or not this is something that they can use.

The standard edition is for the experienced visual observer and would be useful for the pursuit of faint objects. It comes packed with useful tools, including:

- Nightly Planner, for planning and prioritizing observations.
- Nightly Observing List Generator, to create observing lists customized for your conditions, telescope, and experience.

- Sky Events Planner, to look for and plan for special events such as eclipses, occultations, transits, and more.
- Observing Logbook, to collect, organize and display your observations.
- Interactive Atlas, which charts the skies and can be set to your preferences in terms of what is displayed and how it looks.
- Custom Finder Charts for Binoculars and Telescopes, a useful guide to your starhopping to get to even the most difficult objects.
- Naked Eye and Overhead Sky Charts, which takes into account your local light pollution and is an accurate reproduction of your night sky.
- Visual Context Viewer, which projects your telescope field of view on the Interactive Atlas and keeps track of where your telescope is; you can even move the telescope by dragging the FOV circle around on the Interactive Atlas.
- Add Notes, Images, and Web Links to Objects, which helps you make your observing logs truly unique and detailed.

Database Power Search is a very powerful tool to search the extensive *SkyTools* databases. Here are some of the things you can do with this powerful tool:

- Find the most distant quasar visible in your telescope
- Make an observing list of edge-on galaxies
- Find very red stars
- List every dark nebula in a given constellation.

For the observation of faint objects, *Sky Tools* has a lot of useful features. There is a link to the Digital Sky Survey (DSS) image database, where you can download and display images on star charts. This is especially useful if (or maybe we should say "when") you want to make sure that you are really seeing a faint object. You simply right-click on a chart to download a Digital Sky Survey image for display in the chart background or automatically download images for all the objects in your observing list.

For supernova hunts *Sky Tools* is also useful. It can access up to date information on novae and supernovae. The Skyhound website maintains observing lists with currently visible comets, novae, supernovae, and asteroids that make close approaches to Earth. You can easily download these lists from within *SkyTools* so that you will always have up to date information on the transient objects that grace the night.

Professional Edition Features of SkyTools

When you are ready to move on to the next level of observing, whether it is going deeper visually or starting digital imaging of various objects, *SkyTools* has lots of utilities to help in its Professional Edition. There is an Imaging Session Planner that supports deep sky, planetary, lunar, and solar imaging for astronomical CCDs, digital cameras, and video/web cams. This utility helps you to decide when each object is best imaged, what exposure times to try, and how to best stack and process your images. There is a graphic display option that shows the best time to image each of your chosen objects during the night.

SkyTools Pro helps answer the following questions that may come up as you dive into the world (the universe…) of astronomical imaging (these are presented also on the Skyhound.com website):

- How far into twilight should I expose? Which filter is least affected by twilight?
- What will be the difference in my final image if I stack 5-min sub-exposures rather than 30-min subs?
- Should I continue to observe after the Moon rises? If so, which filter is least affected by moonlight?
- What order should I use for my filters?
- How much would I gain by moving to a dark site?
- What effect will poor seeing have on my final image?
- How long can I image Jupiter tonight before the final stacked image will appear smeared from rotation?
- What is the penalty, if any, for imaging Jupiter in daylight?

The program provides answers that help to take the guesswork out of imaging. These come in the form of many utilities, such as the exposure calculator, which can be used to help answer your specific questions with confidence and certainty. There is even a popup window for the Star Atlas that displays a simulated image taken with your camera and telescope, a sort of preview of what you may capture with your imaging system.

SkyTools has a galaxy worth of stars in its database, 522 million down to the 20th magnitude over the entire sky. The data that is available is of the highest quality and includes star information from source catalogs such UCAC and selected stars from USNO-B1.0. Unlike other astronomical software, double and variable star data is fully integrated into the program.

The website www.skyhound.com has this and lots more information on the fine details of *SkyTools 3*. Visit this website for more information and to purchase a copy should you decide to make this a part of your toolkit for your faint objects (or other types of astronomical observing) program.

Appendix E

Astronomical League Observing Clubs

As already mentioned back in Chap. 8, a great place to start observing deep space objects is the Astronomical League (AL). The AL has more than three dozen observing programs that range from outreach and education to the Arp Peculiar Galaxy Club and everywhere in between. A complete listing is at http://www.astroleague.org/al/obsclubs/AlphabeticObservingClubs.html. As websites do change over time, do a web search for the item or items you are looking for if the site has changed. Another step you can take in this specific cases is to go to http://www.astroleague.org and look for the link to the observing clubs (the number of which are growing as new clubs are regularly added).

Also included here is a list of clubs relevant to deep sky and faint object observing, with a brief description of each intermediate and advanced club only. Club descriptions are courtesy of the Astronomical League.

Here are a few suggestions for clubs to pursue depending on your observing experience:

Beginner

If you are just starting out in the fine field of astronomy, you should practice with some of the brighter and easier objects before going deep. The following are excellent with helping you do just that.

- Binocular Double Star Club
- Binocular Messier Club
- Carbon Star Club
- Dark Nebula Club
- Deep Sky Binocular Club
- Double Star Club
- Messier Club
- Southern Skies Binocular Club
- Variable Star Club

B. Cudnik, *Faint Objects and How to Observe Them*, Astronomers' Observing Guides, DOI 10.1007/978-1-4419-6757-2, © Springer Science+Business Media New York 2013

Intermediate

Note: the descriptions are paraphrases of the descriptions that appear on the Astronomical League website, www.astroleague.org, and are used with permission of the personnel who maintain the list and the league.

Caldwell Club

Although Charles Messier cataloged faint fuzzies, and the Herschels recorded many more faint objects, Sir Patrick Caldwell-Moore of the twentieth century has made a catalog of beautiful and interesting objects you should, literally, go out of your way to observe. Two of the obvious objects were never even given NGC designations. Sir Patrick made this list of 109 objects from his home in Selsey, England, using his own observatories. These objects range from magnitude 1 through 13, and Declination +85° to –80°, so some diligence will be required in order to complete this list. There are classified either as bright objects and faint objects (and objects in between). The benefits far outweigh any inconvenience; however, you will be treated to many wonderful new sights to behold in the night sky, and maybe even make some new international observing friends along the way.

Globular Cluster Club

The purpose of this observing club is to introduce the observer to some of the finest globular clusters in the heavens. There are 160 of these objects that are gravitationally bound to the Milky Way and the GC club features the best of these. You will probably benefit from using an 8-in. or larger 'scope, although the club requirements do permit any telescope. This program asks you to observe the structural properties of these diverse celestial objects and compare their sizes, shapes, stellar distributions, and luminosities.

Herschel 400 Club

(The following is quoted from the AL website description for this observing project, http://www.astroleague.org/al/obsclubs/herschel/hers400.html). "For many years, amateur astronomers have enjoyed the challenge and excitement provided by the Messier Club of deep-sky objects. The 110 or so objects in the Messier Catalog introduce the observer to the importance of careful observing and record keeping. Upon completion of this project, however, the amateur was left somewhat in a void. He or she wants to further the quest for deep-sky objects, but outside of the vast New General Catalog, there was no organized program that would provide that next vital step upward. With this idea in mind, the formation of the Herschel Club began".

"It started over 4 years ago, when several members of the Ancient City Astronomy Club in St. Augustine, Florida, who had recently completed the Messier Club noticed a letter in *Sky & Telescope* magazine from James Mullaney of Pittsburgh, Pennsylvania. Mr. Mullaney alerted amateurs to the William Herschel Catalog of deep-sky objects, and suggested this would be a good project to get into after completion of the Messier Lists. He went on to say that Herschel's listings could be found in the original New General Catalog by Johann Dreyer, available from the Royal Astronomical Society in England".

"The New General Catalog was a compilation of several deep-sky catalogs circa 1880; it contained almost 8,000 objects of which 2,477 of these objects were observed by William Herschel. Ancient City Astronomy Club (A.C.A.C.) members began the difficult process of separating his objects, which used a rather unique classification system with eight sub-categories; each individual object was placed into a particular subcategory. These subcategories are:

Class I—Bright Nebulae;
Class II—Faint Nebulae;
Class III—Very Faint Nebulae;
Class IV—Planetary Nebulae;
Class V—Very Large Nebulae;
Class VI—Very Compressed and Rich Clusters of Stars;
Class VII—Compressed Clusters of Small and Large Stars;
Class VIII—Coarsely Scattered Clusters of Stars".

"It was soon discovered that a vast majority of Herschel's objects were in Class II and III, faint and very faint nebulae, with magnitudes fainter than 13, beyond the reach of many amateur telescopes. We of the A.C.A.C. decided that the proposed Herschel Club should consist of enough objects to present a distinct challenge, yet still be within range of amateurs who possessed only modest equipment and were affected by moderate light-pollution problems. After considerable study, we set 400 as the best number of objects to comprise the Herschel Club. Our main references through this process were the *Atlas of the Heavens* and *Atlas of the Heavens Catalog* by Antonin Becvar. These two volumes are readily available to the amateur astronomer and contained all the positions, magnitudes and other pertinent data used in this manual."

The objects listed can be seen in a 6-in. or larger telescope, and these objects include a few Messiers and some of the brighter deep-sky objects from the NGC (including the Double Cluster). There are also many faint objects that qualify for inclusion in the observing lists of this book. For example, the Virgo galaxy field along with the Monoceros Milky Way will present the toughest challenges.

The Herschel 400 project is meant to be an advanced project for those who already have some deep-sky object observing experience. If you are just starting out, do the Messier club first, then do the Herschel 400 club. This is rather like a sequence of college courses, whereby one is a foundational course that serves as a prerequisite course for the next advanced course. Think of the Messier project as being the prerequisite "course" for the Herschel 400 project. These projects truly provide the ultimate astronomical lab experience.

Southern Sky Telescopic Club

For those who get the opportunity to travel to the southern hemisphere, this is an excellent resource to provide a list of the best objects to observe "down under," objects that we cannot see from the northern hemisphere. There are challenges besides being able to travel down south (or if you happen to live in South America, southern Africa, Australia, or New Zealand, to travel up north): there are areas of sky, including the Milky Way, that you have never seen before, new constellations or old ones in unfamiliar orientations, and old favorites that are now high overhead. Some of the objects on this list are objects easily seen with the naked eye but look really spectacular through binoculars or a telescope. One of the obvious requirements to complete this list is you have to actually go there. You cannot observe or image these objects remotely and besides, it is just not the same as actually being there!

Urban Observing Club

This observing club is a project that you literally do not have to leave home to complete, unless you have the good fortune to live under dark skies. The purpose of this club is to bring amateur astronomy back to the cities and suburbs that are affected by significant light pollution (such as your own backyard). Amateur astronomy was once called "backyard astronomy" because to enjoy good views of the universe all one had to do was go into one's backyard, but this expression was coined back in the days when light pollution was not a problem. Since the middle of the twentieth century, skies in the cities and suburbs have brightened significantly, forcing astronomers to make a trip up to 100 miles (or more) into the country to enjoy undimmed views of their favorite objects.

Unfortunately many people do not have the time or the resources to make this trip to get to dark skies, and as the price of gasoline continues to climb in the second decade of the twenty-first century, more and more of us will not be able to afford regular trips to dark sky locales. This is the purpose for the Urban Observing Club, to allow those who stay at home to be able to enjoy the wonders of the universe from just out the back door, and to maximize the observing experience despite heavy light pollution. This list was put together by observers across the nation in urban settings with skies that have limiting magnitudes of 4 down to 2.

Advanced

Again, here is a brief description of each club; the reader can check the AL website for requirements to complete each project as well as the observing lists themselves.

Arp Peculiar Galaxy Club

The Astronomical League has sponsored the Arp Peculiar Galaxy Program for several years. This program comes in two parts: the Northern Arp and the Southern Arp programs. Both groups have the requirement of observing or imaging 100 Arp galaxies from the northern or the southern list. The southern list has 122 galaxies located north of −30° declination, so it is possible for northern hemisphere observers to complete this program. The southern list has a total of 498 galaxies brighter than 15th magnitude for those that are able to observe from the southern hemisphere.

Flat Galaxies Club

The detection of shape and form in galaxies is a visual challenge. As a general class of galaxies, edge-on galaxies are a pleasant yet demanding change of pace in faint object observation. A subclass of edge-on galaxies are flat galaxies that are defined as having a diameter larger than 40 arc-seconds and a major to minor axis ratio greater than or equal to 71. This unique brand of galaxies is used as a tool for studying large-scale motions and distances. Karachentsev et al. published the fact that there is a tight correlation between their linear diameter and the width of the 21 cm neutral hydrogen line that can be detected. There are thousands of flat galaxies cataloged, with many of them well beyond the limit of visual observations. The galaxies that have been selected for this list are brighter than 15th magnitude, with the majority well within the range of amateur telescopes.

Galaxy Groups and Clusters Club

This club features a project that presents a list of 250 galaxy groups and clusters and is designed for detailed visual and/or CCD observation. For many an observer, the challenge of this project will be to actually see these galaxy clusters with their own eyes. Most of the galaxy groups and clusters of this observing program are visually accessible with a 12.5-in. telescope, although there are individual galaxies in some of the groups that are beyond almost all amateur visual observers. Visually, observers will be able to see at least some galaxies in all of the groups.

Working on this challenging project provides several rewards over and above the certificate and pin that come with successful completion of the list. For example, subtle visual detection will reward you with the thrilling experience of viewing extremely distant objects at the very edge of visibility. In a CCD image, this same cluster may be a very beautiful cloud of galaxies and their dim clouds of globular clusters. One of the challenges of this program is to match galaxies with their map coordinates, which requires the ability to identify star fields. Many of the clusters in this observing project's catalog are quite faint and difficult to find and will require the observer to employ many of the techniques described in this book.

Herschel II Club

This project can be viewed as a continuation of the Herschel 400 program. The list used for this club includes 400 of the 2,478 deep sky objects cataloged by William Herschel in the late 1700s and represents the next level of observing after the original Herschel 400 program. The Herschel II observing program is an advanced level project focused on improving observers' technical skills by taking thorough field notes and developing accurate technical object descriptions.

The 400 objects in this program include 323 galaxies, 41 open clusters, 21 nebulae, 9 planetary nebulae, 3 cluster-nebulae, and 3 globular clusters. Most of the objects are between magnitude 11 and 13. One can observe many of the objects with an 8-in. telescope under transparent dark skies, but a 10-in. telescope (or larger) will probably be necessary to complete the list. A major challenge of this program is the fact that many of the objects on the list are in sky fields that contain numerous other objects, some even superimposed on top of each other. To successfully complete this program the observer will have to be able to pick out and identify the correct object from the many objects that he/she will find under these circumstances.

Local Galaxy Groups and Neighborhood Club

The Local Group of galaxies is dominated by three large spiral/barred spiral galaxies: the Andromeda Galaxy (M31), the Triangulum Galaxy (M33), and the Milky Way Galaxy. There are many dwarf galaxies that orbit these three major galaxies. Most of these faint dwarf galaxies are not as easily seen as their brighter companions. Investigations from various radio telescopes from around the world have discovered even more companion galaxies around the Milky Way.

The Local Group of galaxies is gravitationally bound in a cloud surrounding the more massive galaxies. More and more diffuse clouds of stars associated with the Local Group are being discovered by professional astronomers who have also found streams of stars within our own Milky Way that come from dwarf galaxies being cannibalized by our massive home galaxy. Some of these complexes in the Local Group and Galactic Neighborhood are observable in modest-sized telescopes. As a consequence, amateur astronomers can view the brighter representatives of these scientifically significant objects, many of which are in popular catalogs, such as the Messier and Caldwell.

Open Cluster Club

Open clusters are of tremendous importance to the science of astronomy, if not to astrophysics and cosmology in general. These beautiful objects serve as the "laboratories" of stellar astronomy, since all of the member stars are at nearly the same distance and were all born at essentially the same time. Each cluster is a running experiment, where we can observe the effects of composition, age, and environment on the evolution of their members. Since we can only see a snapshot in the evolution of a single cluster, we can look at many of them of various ages and in various stages of evolution to piece together a picture of stellar evolution in general.

Not only do open star clusters make great astrophysical laboratories, they also serve as important tracers of the Milky Way and other parent galaxies. Serving as interstellar probes, these objects help us understand the current structure of galaxies and gain some insight into the creation and evolution of galaxies. Starting from two of the nearest and brightest clusters, the Hyades and the Pleiades, and then going to more distant clusters, open clusters serve to define the distance scale of the Milky Way, and from there all other galaxies on out to the entire visible universe.

Anyone who has looked at a cluster through a telescope or binoculars has realized that these are objects of immense beauty and symmetry. Whether a cluster like the Pleiades, seen with delicate beauty with the unaided eye or in a small telescope or binoculars, or a cluster like NGC 7789, whose thousands of stars are seen with overpowering wonder in a large telescope, open clusters bring awe and amazement to the observer.

The best of all is that these wondrous sights are available to everyone who would like to have a look. Whether one uses a large or small telescope, or observes from a dark sky or light-polluted location, clusters are there waiting on any clear night for us to take a look. After completing this club, it is hoped that the observer will come away from this experience with a new appreciation to the wonder, uniqueness, and beauty of each open star cluster.

Planetary Nebula Club

Planetary nebulae are perhaps the most interesting and beautiful objects in the heavens. These display a wonderful range of variation, and no two of them are exactly alike, sort of like snowflakes on Earth. Planetary nebulae show a vast array of complex shapes and (in some cases) vibrant colors. The study of these objects is essential to understanding one of the endpoints of stellar evolution, as sun-like stars end their lives in this manner.

Pursuing this observing club should instill in the observer a new and fresh appreciation for these ghostly and beautiful objects. There are a least 1,000 of these objects visible in amateur-sized telescopes, and many splendid examples are not on the observing list for this club, but you can look for these on your own, either as you are working your way through the list provided by this club or afterwards. Take time to stare long at these gems and consider their interesting natures.

Appendix F

How to Find More Observing Lists of Faint Objects

As already stated several times throughout this book, there are literally gobs of projects out there for those interested in tracking down as many faint objects as they can get their eyes on, and this section provides a sampling of such lists. These are all on the Internet and should be around for some time to come.

The first ten lists here provide links to additional projects, lists, and information that came with many of the tables presented in Chap 9. The eleventh is a re-statement (or rewritten) of the list of projects mentioned in Chap. 8, but with a bit more detail. The last item provides a link to a visual survey of the Magellanic Clouds. The author of this relates her experience in surveying these, the two largest satellite galaxies of the Milky Way, and one can read about this, and even get a list and information about objects to hunt for, in the link provided (assuming the observer lives in or will visit the southern hemisphere, which is necessary in order to view these irregular satellite galaxies of the Milky Way).

Sharpless HII Regions

- Adventures in Deep Space—Sharpless HII http://www.astronomy-mall.com/Adventures.In.Deep.Space/sharp.htm
- Sharpless Images—http://galaxymap.org/cgi-bin/sharpless.py?s=1

Extragalactic Globulars

- Hodge Atlas of the Andromeda Galaxy—http://ned.ipac.caltech.edu/level5/ANDROMEDA_Atlas/frames.html
- Observations of M31 Globulars—http://www.astronomy-mall.com/Adventures.In.Deep.Space/gcm31.htm

B. Cudnik, *Faint Objects and How to Observe Them*, Astronomers' Observing Guides, DOI 10.1007/978-1-4419-6757-2, © Springer Science+Business Media New York 2013

- Extragalactic Globulars—http://www.astronomy-mall.com/Adventures.In.Deep.Space/gcextra.htm
- Observations of M33 Globulars—http://www.astronomy-mall.com/Adventures.In.Deep.Space/gcm33.htm

Palomar/Obscured Globulars

- Observations of Palomar Globulars—http://www.astronomy-mall.com/Adventures.In.Deep.Space/palglob.htm
- Palomar Globs—Images and Sketches—http://www.deepsky-visuell.de/Projekte/PalomarGC_E.htm
- Observations of Obscure Globulars—http://astronomy-mall.com/Adventures.In.Deep.Space/obscure.htm

Local Group Dwarf Galaxies

- SEDS page on Local Group Galaxies—http://messier.seds.org/more/local.html
- Barnard's Galaxy in depth—http://www.astronomy-mall.com/Adventures.In.Deep.Space/barnard.htm

Compact Galaxy Groups

- Observations of Interesting Hickson Groups—http://www.astronomy-mall.com/Adventures.In.Deep.Space/hicklist.htm (with finder charts)
- Observations of Shakhbazian Compact Groups of Compact Galaxies—http://www.andreas-domenico.de/astro/shkh.html

Arp Interaction Pairs

- http://nedwww.ipac.caltech.edu/level5/Arp/frames.html
- Martin Schoenball—images, sketches and observing notes—http://arp.schoenball.de/index_e.htm
- Arp Sampler with notes from Steve Gottlieb—http://www.astronomy-mall.com/Adventures.In.Deep.Space/arpsampler.html

Abell Galaxy Clusters

- Best of the Abell Galaxy Clusters from Adventures in Deep Space—http://www.astronomy-mall.com/Adventures.In.Deep.Space/agctable.html
- Saguaro Astronomy List and observations of Abell Clusters—http://www.saguaroastro.org/content/Abel-Galaxy-Clusters.html
- Andreas Domenico survey of Abell Galaxy Clusters—http://www.andreas-domenico.de/astro/agc.html
- Albert Highe's survey of Abell 426—http://pw2.netcom.com/~ahighe/a426.html

HII Regions in External Galaxies

- HII regions in Barnard's Galaxy—http://www.astronomy-mall.com/Adventures. In.Deep.Space/barnard.htm

Ancient Giant Planetaries

- Abell Planetaries—lists, images, notes from Steve Gottlieb—http://www.astronomy-mall.com/Adventures.In.Deep.Space/abellpn.htm
- Eric Honeycutt's Observations of Abell PN—http://www.stathis-firstlight.de/deepsky/abell_honeycutt.htm

Quasars

- Catalog of Quasars—http://www.klima-luft.de/steinicke/index_e.htm
- Observing the Double Quasar—http://www.astronomy-mall.com/Adventures. In.Deep.Space/dblqso.htm

Listed Already but with More Details

- 100 Peculiar Galaxies
- Off the Beaten Path (objects organized by season)
- Observing Galaxy Clusters (there are links to images of 16 clusters and super-clusters of galaxies, a handful of which are provided in the next chapter)
- The Abell Planetaries (86 gems to test your observing skills)
- The Ultimate Observing List
- Steve Gottlieb's NGC Notes
- Catalogs, Lists and Links
  Abell Galaxy Clusters
  GSSP General List (summer—html)
  Lynds Dark Nebulae
  Abell Planetary Nebulae
  Herschel 400-I
- M31 Globular Catalog
- Arp Catalog
- Herschel 400-II
- M33 HII Regions and Star Clouds
- Barnard Dark Objects
- Herschel 2500
- Off the Deep End (10 challenges)
- Orion Deep Map 600 (Megastar file)
- Hickson Galaxy Clusters
- Shakhbazian Galaxy Groups
- Galaxy Trios
- Lynds Bright Nebulae
- Sharpless HII Regions

- GSSP Challenge List (summer—html)
- Detailed monthly observing lists from February 2007 to January 2010
- Observing Reports

Texas Star Party Favorites

- http://texasstarparty.org/activities/tsp-observing-programs/ provides links to pages that introduce and provide links to lists of current year observing programs
- Regular telescope observing program archives (up to last year)→http://texasstarparty.org/activities/tsp-observing-programs/tsp-observing-program-archive/
- Advanced telescope observing programs (up to last year)→http://texasstarparty.org/activities/tsp-observing-programs/tsp-observing-program-archive/

A Visual Atlas of the Magellanic Clouds and More!

- http://www.webbdeepsky.com/ has a plethora of resources
- http://www.webbdeepsky.com/wperiodical/VisualAtlas.html is a link to information on one such resource: a tour of the Magellanic Clouds, written by Jenni Kay, Southern Sky director of the Webb Deep Sky Society.

Index

B. Cudnik, *Faint Objects and How to Observe Them*, Astronomers' Observing Guides,
DOI 10.1007/978-1-4419-6757-2, © Springer Science+Business Media New York 2013